Springer Series in **Materials Science** 1

Edited by Hans-Joachim Queisser

Springer Series in **Materials Science**
___

Editors: Aaram Mooradian   Morton B. Panish

Dieter Bäuerle

Chemical Processing with Lasers

With 88 Figures

Springer-Verlag Berlin Heidelberg New York
London Paris Tokyo

Professor Dr. *Dieter Bäuerle*

Angewandte Physik, Johannes-Kepler-Universität Linz,
A-4040 Linz, Austria

TA
1677
·B38
1986
SCIENCE
LIBRARY

Guest Editor:

Professor Dr. *Hans-Joachim Queisser*

Max-Planck-Institut für Festkörperforschung, Heisenbergstraße 1,
D-7000 Stuttgart 80, Fed. Rep. of Germany

Series Editors:

Dr. *Aaram Mooradian*

Leader of the Quantum Electronics Group, MIT,
Lincoln Laboratory, P.O. Box 73, Lexington, MA 02173, USA

Dr. *Morton B. Panish*

AT&T Bell Laboratories, 600 Mountain Avenue, Murray Hill, NJ 07974, USA

ISBN 3-540-17147-9 Springer-Verlag Berlin Heidelberg New York
ISBN 0-387-17147-9 Springer-Verlag New York Berlin Heidelberg

Library of Congress Cataloging-in-Publication Data. Bäuerle, D. (Dieter), 1940-. Chemical processing with lasers. (Springer series in materials science ; v. 1). Bibliography: p. Includes index. 1. Lasers–Industrial applications. 2. Materials–Effect of radiation. I. Title. II. Series. TA1677.B38 1986   621.36'6   86-26203

© Springer-Verlag Berlin Heidelberg 1986
Printed in Germany

The use of registered names, trademarks, etc. in this publication does not imply, even in the absence of a specific statement, that such names are exempt from the relevant protective laws and regulations and therefore free for general use.

Offset printing: Druckhaus Beltz, 6944 Hemsbach/Bergstr. Bookbinding: J. Schäffer OHG, 6718 Grünstadt
2153/3150-543210

Preface

Materials processing with lasers is a rapidly expanding field which is increasingly captivating the attention of scientists, engineers and manufacturers alike. The aspect of most interest to scientists is provided by the basic interaction mechanisms between the intense light of a laser and materials exposed to a chemically reactive or nonreactive surrounding medium. Engineers and manufacturers see in the laser a new tool which will not only make manufacturing cheaper, faster, cleaner and more accurate but which also opens up entirely new technologies and manufacturing methods that are simply not available using existing techniques. Actual and potential applications range from laser machining to laser-induced materials transformation, coating, patterning, etc., opening up the prospect of exciting new processing methods for micromechanics, metallurgy, integrated optics, semiconductor manufacture and chemical engineering.

This book concentrates on the new and interdisciplinary field of laser-induced *chemical* processing of materials. The technique permits maskless single-step deposition of thin films of metals, semiconductors or insulators with lateral dimensions ranging from a few tenths of a micrometer up to several centimeters. Moreover, materials removal or synthesis, or surface modifications, such as oxidation, nitridation, reduction, metallization and doping, are also possible within similar dimensions. This book is meant as an introduction. It attempts to cater for the very broad range of specific interests which different groups of readers will have, and this thinking underlies the way in which the material has been arranged. Some chapters are intended to give a deeper and more general insight into the fundamental mechanisms and discuss these on the basis of model systems. The extended tables are included so as to give the reader a quick and comprehensive overview of the "state of the art" and to refer the engineer or manufacturer on to the original literature of his particular area of interest as quickly as possible.

Clearly, there is no way in which a book on such a new and rapidly developing field can possibly be complete, either with respect to the interpretation and comprehension of results or with respect to the technique's many limitations and possibilities - the latter being frequently a matter for speculation in any case. The author would be grateful for suggestions and criticism which could improve and complete any forthcoming editions of this book.

I wish to thank Dr. H.K.V. Lotsch of Springer-Verlag for the suggestion that I should write this book, and to record my very great thanks to all my staff, colleagues and friends - in particular to Dr. R. Kullmer, Dipl.-Ing. K. Piglmayer and Prof. U.M. Titulaer - for valuable discussions and for their critical reading of various parts of the manuscript. I am deeply indebted to my assistant and secretary, Dipl.-Ing. Irmengard Haslinger, for her tireless assistance in writing this book. Last but not least, I wish to express my deep gratitude to my family: Barbara, Anne Friederike and Christoph Fridemann for their constant encouragement and support. This book is dedicated to the memory of my esteemed father Karl Wilhelm Bäuerle.

Linz, May 1986   **Dieter Bäuerle**

Contents

1. Introduction

The extent and the intensity of current interest in the use of lasers, be it for scientific investigations or for industrial applications, is directly linked to the inherent properties of laser light: its coherence and its monochromaticity. Its spatial coherence permits extreme focusing and directional irradiation at high energy densities. The monochromaticity of laser light, together with its tunability, opens up the possibility of highly selective narrow-band excitation. Controlled pulsed excitation offers high temporal resolution and often makes it possible to overcome competing dissipative mechanisms within the particular system under investigation. The combination of all of these properties offers a wide and versatile range of quite different applications. The fundamentals regarding the generation and properties of laser light and the various types of lasers have already been extensively reviewed (see, for example, [1.1-5]).

Materials processing with lasers takes advantage of virtually all the properties of laser light: The high energy density and directionality of laser light permits strongly localized heat and/or chemical treatment of materials, with a resolution down to less than 1 μm. Pulsed lasers or scanned cw lasers allow time controlled processing between about 10^{-14} s and continuous operation. The monochromaticity of laser light allows for control of the depth of heat treatment and/or selective, nonthermal excitation - either within the surface of the material or within the molecules of the surrounding medium - simply by tuning the laser wavelength. Because laser light is an essentially massless tool, it does not bring any appreciable mechanical force to bear on the material being processed, thereby obviating the need for mechanical holders with all the attendant problems these pose in the case of brittle materials. This same factor makes the laser beam capable of being moved at speeds which can never be obtained using mechanical tools or conventional heat sources. Contrary to mechanical tools, laser light is not subject to wear and tear. This avoids any contamination of the material being processed, and, if the beam is properly controlled, also guarantees constant processing characteristics. Laser technology is, of course, completely compatible with present-day electronic control techniques. Naturally, a particular processing application will require only one or a few of these properties.

Materials processing with lasers can be classified into *conventional* chemically nonreactive processing and *chemical* processing. In the following we shall briefly introduce and define these two fields.

Conventional laser processing is essentially heat treatment that can be performed in a vacuum or in a chemically nonreactive atmosphere. The temperature distribution induced by the absorbed laser light depends on the absorption and the thermal diffusivity of the processed material and, near phase transitions, on transformation energies for crystallization, melting, boiling, etc. The best-established processing application is laser machining such as drilling, scribing, cutting, welding, bonding, trimming and shaping of materials. Here, the beam of a high power laser, such as a CO_2 or Nd:YAG laser, is focused onto the material to induce local melting and/or vaporization ⌊1.6-14⌋. Typical light intensities used range from about 10^5 to 10^9 W/cm^2.

Alongside these well-established applications, laser-induced structural transformations in material surfaces and thin films are attracting increasing interest. Among these are surface hardening, cladding and alloying (in the sense of freezing a nonequilibrium state) ⌊1.6-16⌋, and also defect annealing, recrystallization and amorphization of materials surfaces ⌊1.6, 16-30⌋. Besides the laser intensity, one of the most important parameters in these applications is the laser beam dwell time, i.e. the average time of laser light illumination at a particular site. The much lower energy densities necessary to induce structural transformations (typically 10^3 - 10^5 W/cm^2) also make it possible to process larger areas of material simultaneously, by using unfocused or defocused laser beams. Typical lateral dimensions of the processed area range from several centimeters (for a single scan) to less than one micrometer. The depth of penetration into the material surface ranges from some tens of Ångströms to several centimeters, depending on the optical properties of the material and the laser wavelength.

Laser-induced transformation hardening of surfaces is generally performed at intensities that do not result in surface melting. Laser cladding and alloying, on the other hand, require, in general, melting of the solid surface. In recent years considerable emphasis has been placed upon laser annealing of ion-implanted semiconductor surfaces, and in particular upon that of ion-implanted Si. Here, the rapid heating and cooling rates achieved

2

with lasers make it possible to avoid dopant redistribution. Recrystallization of amorphous films is being investigated with the intention of producing cheaper films. Again, Si is one of the most intensely studied materials.

From a chemical point of view, conventional laser processing does not require activation of any chemical reaction. This is quite different in laser-induced chemical materials processing. The object of laser-induced chemical processing (LCP) of materials is the patterning, coating and physicochemical modification of solid surfaces by activation of real chemical reactions. Here, the laser-induced activation or enhancement of a reaction may be based on *pyrolytical* (photothermal) and/or *photolytical* (photo-chemical) excitation mechanisms. Laser light may induce or enhance reactions either heterogeneously in adsorbate-adsorbent systems, at gas-solid or liquid-solid interfaces, within solid surfaces itself, or homogeneously within surrounding media near solid surfaces. Such reactions may result in materials deposition, surface modification (oxidation, nitridation, reduction, metallization, doping, etc.), materials synthesis, or removal of material ⌊1.6,26-44⌋. As in conventional processing, chemical processing can be performed locally or on a more extended scale.

Local processing allows single-step *direct* writing of patterns with lateral dimensions down into the submicrometer range. Today, such patterns are produced by large-area processing techniques, e.g. conventional chemical vapor deposition (CVD) or plasma processing, together with mechanical masking or lithographic methods. Contrary to laser direct writing, these standard techniques require several production steps. Furthermore, the nonlinearity of laser-induced chemical reactions makes it possible to increase the resolution over that in standard photolithography. In other words, it is possible to produce structures with lateral dimensions smaller than the diffraction-limited diameter of the laser focus.

Large-area chemical processing can be performed either with laser light propagating perpendicular to the substrate surface (normal incidence) or, additionally, with laser light propagating parallel to the substrate surface. The latter technique allows extended thin film processing with or without uniform heating of the substrate.

In any optical configuration, LCP differs significantly from conventional techniques. Contrary to standard high temperature techniques (for example CVD), LCP makes it possible to restrict or even virtually eliminate heat treatment of the material being processed. Consequently, the laser technique also allows one to process temperature-sensitive materials such as compound

3

semiconductors (GaAs, InP, etc.) and polymer films. Furthermore, LCP avoids the damage to materials - from ion or electron bombardment or from overall vacuum ultraviolet radiation - which is inherent to conventional low-temperature techniques such as plasma processing and ion- or electron-beam processing. Laser processing is, of course, not limited to planar substrates and allows three-dimensional fabrication as well.

The present book concentrates on the new field of laser-induced chemical processing of materials as defined just above. As already mentioned in the Preface, the intention is to give scientists, engineers and manufacturers an overview of the extent to which the technique is understood at present, and of the various possibilities and limitations it has.

Determining the scope of this book was, naturally, very difficult and in some cases somewhat arbitrary. For example, we have paid little or no attention to problems related to laser machining taking place in air, where oxidation or nitridation reactions may play an important role as well. Furthermore, laser-induced structural transformations, or laser alloying, are often accompanied by a precipitation of new compounds within the surface of the material, and such cases could equally well be incorporated into the book. We also exclude the wide field of homogeneous laser photochemistry. Here, the basic interaction mechanisms between molecules and laser light are investigated, and materials are produced homogeneously in the gas or liquid phase ⌊1.45-50⌋. Sometimes such reactions also take place in the presence of solid surfaces that catalyze the laser-induced reaction. Finally, no medical applications have been included, even though these tend to be based on mainly photothermal or photochemical reactions, in the same way as the cases discussed in this book ⌊1.51-55⌋.

The book is divided into nine chapters. Chapters 2 and 3 are devoted to the fundamental mechanisms in pyrolytic and photolytic chemical processing and to the kinetics of laser-induced chemical reactions at or near solid surfaces. Experimental techniques are briefly described in Chap.4. Chapter 5 outlines laser-assisted deposition of materials in the form of microstructures and thin extended films. Laser-induced surface modifications are described in Chap.6. The formation of stoichiometric compounds is covered in Chap.7. Chapter 8 deals with laser-induced chemical etching of materials. In Chap.9 actual and potential applications of LCP in micromechanics, microelectronis, integrated optics and chemical technology are summarized. The main abbreviations and symbols used throughout are listed at the end of the book.

2. Fundamental Excitation Mechanisms

Laser-induced chemical reactions can be based on several fundamentally different microscopic mechanisms. In the following we shall classify reactions into those which are governed by *mainly* pyrolytic (photothermal) or by *mainly* photolytic (photochemical) processes. We shall call a reaction pyrolytic if the thermalization of the laser excitation is fast compared to the reaction, and photolytic if this is not the case, i.e. when the constituents of the reaction are in nonequilibrium states. The laser excitation can take place within the ambient gaseous or liquid medium and/or directly within the surface of the solid material (substrate) to be processed. In many cases the different mechanisms and possibilities of excitation contribute simultaneously to the reaction, but often one of them dominates. There are also many examples where, for example, a reaction is initiated photolytically and proceeds pyrolytically, or vice versa.

   In this chapter, we will outline a phenomenological treatment of laser-induced temperature distributions for semiinfinite substrates (Sect.2.1.1) and compare it to model structures that are relevant in LCP (Sect.2.1.2). Sections 2.2.1 and 2.2.2 deal with dissociative single-photon and multiphoton electronic excitations and with the dissociation of molecules by photosensitization. The fundamentals of selective infrared vibrational excitations are outlined in Sect.2.2.3.

2.1 Pyrolytic Processing

In pyrolytic laser-induced chemical processing, the laser serves as a heat source and gives rise to thermochemistry. The absorbed laser light may heat either the substrate, or the ambient gaseous or liquid medium, or both. High temperatures are reached, e.g. by one-photon absorption processes, if the frequency of the laser light matches a strong vibrational absorption in a particular molecule, or any optically active elementary excitation in the solid. In solids or in single, isolated polyatomic molecules, this excitation energy is randomized, in general, within picoseconds. Because of the high molecular densities used in LCP, randomization of energy *between* single molecules occurs via collisional energy transfer, typically within 10^{-12} - 10^{-7} s. In spite of their thermal character, such laser-driven reactions may be very different from those traditionally initiated by a conventional heat source in which an equivalent amount of thermal energy is deposited. This is because of the much higher temperatures ($\sim 10^4$ K) obtainable in the small reaction volume defined by a focused laser beam. Therefore, novel reaction products due to different reaction pathways may occur.

Single-step production of spatially well-defined structures is usually performed by local substrate heating using laser light that is *not* absorbed by the ambient gaseous or liquid medium. In this case the chemical reaction is essentially confined to the hot spot that is induced on the substrate by the absorbed laser light. Because of the large variety of optically active elementary excitations in solids, and because of the rapid dissipation of the excitation energy, it is clear that in the case of local substrate heating the dependence of the reaction rate on the frequency of the laser radiation is less pronounced than in the case of gas-phase heating mentioned above.

A quantitative analysis of pyrolytic processing based on local substrate heating requires a detailed knowledge of the laser-induced temperature distributions. Direct temperature measurements have been performed with a reliable degree of accuracy in only a very few cases. In many cases the measurement techniques (Sect.4.4) cannot be satisfactorily applied for practical reasons, and so a method of calculating the laser-induced temperatures is desirable. In fact, many features in pyrolytic processing can be qualitatively and in some cases even quantitatively understood from calculated temperature distributions.

The temperature distribution induced by the absorbed laser light depends on the absorption coefficient $\alpha(\lambda,T)$ within the processed area, on the transport of heat, which is determined by the thermal diffusivity $D_t(T)$, on chemical reaction energies (exothermal or endothermal), and on transformation energies for crystallization, melting, boiling, etc., where relevant. Here, λ is the laser wavelength and $T(r,t)$ the laser-induced temperature distribution, which is a function of both the distance r from the center of the beam and of the time t. Note, however, that in laser-induced chemical processing, the quantities α and D_t do *not* refer just to the substrate material. In fact, these quantities are strongly changed by the deposition, surface modification, compound formation or etching process itself [2.1-6]. Let us consider this, for example, in the case of laser-induced deposition

6

from the gas phase when the substrate and the laser beam are static. Before nucleation takes place, α and D_t refer to the substrate. However, when nucleation commences, these quantities will change rapidly with the density of the nuclei and therefore with time. When a compact film is formed, e.g. a metal film, and if the penetration depth of the laser light is smaller than the film thickness, α will refer only to this deposited film. Similarly, the thermal conductivity will be quite different for such a combined structure and for a simple homogeneous plane substrate. The situation is very similar in laser-induced compound formation and etching.

We will now discuss these problems in greater detail and compare laser-induced temperature distributions on semiinfinite substrates with those induced on structures that are relevant in laser-induced chemical processing.

2.1.1 Semiinfinite Substrates

Temperature distributions induced on semiinfinite substrates have been calculated for static and scanned cw [2.1,2,5,7-18] and pulsed [2.12-16,23-26] laser beams. Similar calculations have also been performed for semiinfinite multilayer structures [2.27,28]. Here, in many cases, analytic solutions are possible. In the following we will outline some of the essential features for cw laser and pulsed laser heating of solids. Continuous-wave laser heating is the predominant mode in laser-induced microchemical processing, while pulsed laser heating is mainly applied for large-area chemical processing and for projection patterning (Sect.4.2).

In the following analysis we shall consider the flow of heat by conduction into the bulk of a material irradiated by a slowly moving laser beam. Heat losses into the surrounding medium by thermal conduction, convection, or blackbody radiation are not taken into consideration. The temperature distribution induced by the absorbed laser light can be calculated from the heat equation, which can be written as

$$\frac{\kappa(T)}{D_t(T)} \frac{\partial T}{\partial t} - \nabla\lfloor\kappa(T)\nabla T\rfloor = Q \ . \tag{2.1}$$

Here, $\kappa(T)$ and $D_t(T)$ are the temperature-dependent thermal conductivity and thermal diffusivity of the irradiated material. The thermal diffusivity is given by $D_t(T) = \kappa(T)/\rho_m c_p$, where ρ_m is the mass density and c_p the specific heat at constant pressure. The term Q is the source term arising from the incident laser light. The temperature-dependent thermal conductivity $\kappa(T)$ can be eliminated from the heat equation by performing a Kirchhoff transform [2.29]. The Kirchhoff transform requires the introduction of a linearized

7

temperature Θ, which is defined as

$$\Theta(T) = \Theta(T_0) + \int_{T_0}^{T} \frac{\kappa(T')}{\kappa(T_0)} \, dT' \, , \tag{2.2}$$

where $\Theta(T_0)$ is a constant, and T_0 the temperature at infinity. In terms of the linearized temperature, the heat equation can be written as

$$\frac{1}{D_t(T(\Theta))} \frac{\partial \Theta}{\partial t} - \nabla^2 \Theta = \frac{Q}{\kappa(T_0)} \, . \tag{2.3}$$

In a reference frame that is moving with the laser beam, say in x-direction, and in which the temperature distribution is stationary, the heat equation becomes

$$\frac{v_s}{D_t(T(\Theta))} \frac{\partial \Theta}{\partial x} - \nabla^2 \Theta = \frac{Q}{\kappa(T_0)} \, , \tag{2.4}$$

where v_s is the velocity of the substrate relative to the reference frame. For a Gaussian beam at normal incidence to the substrate, Q is given by

$$Q = \frac{2P(1-R)}{\pi w_0^2} \exp(-2r^2/w_0^2) f(z) \quad \text{with} \tag{2.5}$$

$$f(z) = \alpha \exp(-\alpha z) \, , \tag{2.6}$$

if α is constant. P is the incident laser power, R the reflectivity of the substrate, and $2w_0$ the diameter of the laser focus (Sect.4.2). In the following we shall discuss various different approximations.

a) *High Absorption*

In the case of high absorption ($\alpha > 10^4$ cm^{-1}) with *no* appreciable light penetration into the material, the source term Q vanishes except at the irradiated surface, where it is given by the effective absorbed irradiance. This assumption holds, for example, for metals, and in good approximation also for Si for visible Ar$^+$ or Kr$^+$ laser light, especially at elevated temperatures (see below). In this case, we can write

$$f(z) = \delta(z) \, . \tag{2.6'}$$

Using the source term (2.5) together with (2.6'), the general solution of

8

(2.4) is obtained by a Green's function method [2.29]. It is given by

$$\Theta = \frac{2^{1/2}\, P(1-R)}{\pi^{3/2}\, \kappa(T_0) w_0} \int_0^\infty g(u)\, du \;, \qquad \text{where} \tag{2.7}$$

$$g(u) = (1 + u^2)^{-1} \exp\left\{-2\left[\frac{(X + V\chi u^2)^2 + Y^2}{1 + u^2} + \frac{Z^2}{u^2}\right]\right\} \qquad \text{with} \tag{2.8}$$

$$X = x/w_0, \; Y = y/w_0, \; Z = z/w_0,$$
$$V = v_s/w_0, \; \chi(T) = w_0^2/[8(2^{1/2})D_t]. \tag{2.9}$$

Because $T = T(\Theta)$, and because R and D_t are temperature dependent for most materials, (2.7) is an implicit equation for the linearized temperature Θ. The above treatment cannot be used, however, when D_t depends very strongly on temperature and when phase changes occur. In this case, a more complex formulation of the problem based on numerical methods is required.

We now compare calculated temperature distributions, still assuming $f(z) = \delta(z)$, for various different cases.

The Static Case

In the static case, i.e. for cw laser irradiation and $v_s = 0$, equation (2.7) no longer depends on the diffusivity D_t.

In the first approximation, we also neglect the temperature dependences of parameters. The assumption of a temperature independent thermal conductivity

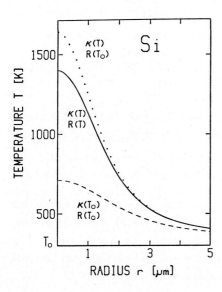

Fig.2.1. Temperature distribution induced by a static Gaussian laser beam at normal incidence. P = 0.575 W, $2w_0$ = 5 μm, T_0 = 300 K. The radius is measured from the center of the laser beam. The parameters correspond to Si. Dashed curve: $\kappa = \kappa(Si,T_0)$, R = R(Si,T_0). Dotted curve: $\kappa = \kappa(Si,T)$, R = R(Si,T_0). Full curve: $\kappa = \kappa(Si,T)$, R = R(Si,T) (after [2.5])

applies to metals at temperatures much greater than the Debye temperature. This follows from the Wiedemann-Franz law $\kappa = \pi^2 k_B^2 T\sigma/3e^2$ and the temperature dependence of the high temperature electrical conductivity $\sigma \sim T^{-1}$ [2.30]. The dashed curve in Fig.2.1, which is calculated from (2.7), applies to this case (note that for $\kappa = \mathrm{const}(T)$ the temperatures Θ and T are equal). In order to compare these results with those presented below, we have chosen values for the parameters which correspond to those of crystalline Si at room temperature. The maximum temperature rise in the center of the spot is

$$\Delta T_c = T_c - T_0 = \frac{P(1-R)}{(2\pi)^{1/2}\,\kappa(T_0)w_0}\;.\qquad\qquad (2.10)$$

Next, we take into account the temperature dependence of the thermal conductivity. For crystalline semiconductors and insulators, the high temperature thermal conductivity can be described, in many cases, by $\kappa \sim T^{-1}$. For Si, a fit to experimental data yields [2.31]

$$\kappa(T) = k(T - T_k)^{-1};\quad k = 299\ \mathrm{W/cm},\ T_k = 99\ \mathrm{K}.\qquad (2.11)$$

With (2.2) and (2.7) the temperature can then be expressed analytically by

$$T(\Theta) = T_k + (T_0 - T_k)\,\exp[\Theta\,/(T_0 - T_k)].\qquad\qquad (2.12)$$

The dotted curve in Fig.2.1 is calculated for this case.

Finally, we also consider the temperature dependence of the reflectivity. For Si, this temperature dependence can be approximated between room temperature and its melting point (1690 K) [2.18-22] by

$$R(T) = 0.324 + 4\cdot 10^{-5}\ T \qquad\qquad (T < 1000\ \mathrm{K})\qquad (2.13a)$$
$$= 0.584 - 4.8\cdot 10^{-4}\ T + 2.6\cdot 10^{-7}\ T^2 \quad (T > 1000\ \mathrm{K})\ .\quad (2.13b)$$

The full curve in Fig.2.1 applies to this case. The increase in reflectivity with temperature reduces the temperature at the center of the irradiated zone by approximately 13%.

The Dynamic Case

We are now interested in the influence of the *scanning* velocity on the temperature distribution. This can be calculated from (2.7) and (2.2) as long as D_t is independent or only slightly dependent on temperature. This assumption is invalid for Si, for which the thermal diffusivity depends strongly on temperature and can be approximated by [2.31]

10

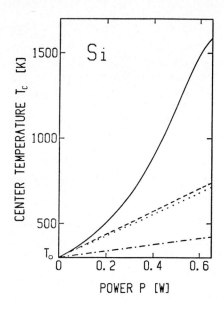

CENTER TEMPERATURE T_c [K]

Si

POWER P [W]

Fig.2.2. Laser-induced center temperature T_c as a function of laser power for different scanning velocities. $2w_0 = 5$ µm, $T_0 = 300$ K. Dashed curve: $v_s = 0$. Dotted curve: $v_s = 1$ m/s. Dash-dotted curve: $v_s = 100$ m/s. In all three cases $\kappa = \kappa(Si,T_0)$, $R = R(Si,T_0)$, $D_t = D_t(Si,1680$ K$)$. The full curve was calculated for $v_s = 0$ and $\kappa = \kappa(Si,T)$ and $R = R(Si,T)$ (after [2.5])

$$D_t(T) = d(T - T_d)^{-1}; \quad d = 128 \ cm^2K/s, \ T_d = 159 \ K . \tag{2.14}$$

In such a case, a numerical solution of the heat equation must be employed. For simplicity, however, we shall confine ourselves to an upper estimate and take the value for the diffusivity near the melting point of Si, i.e. $D_t = D_t(Si,1680$ K$)$, instead of (2.14). Figure 2.2 shows the center temperature as a function of laser power for different scanning velocities v_s (dashed, dotted and dash-dotted curves). It becomes evident from the figure that in solid-phase processing such as laser annealing [2.33-40] or laser synthesis (Chap.7), where scanning velocities of up to several meters per second are common, the influence of the scanning velocity on the center temperature must be taken into account. On the other hand, in processing cases where $v_s < 10^4$ µm/s, the center temperature is essentially unaffected by v_s. The latter approximation also holds in direct writing of patterns by LCVD, if $\kappa_D/\kappa_S \approx 1$ (Sect.2.1.2). The full curve in Fig.2.2 shows the dependence of the center temperature for $v_s = 0$, but for temperature-dependent $\kappa(T)$ and $R(T)$.

For *pulsed* laser irradiation, an approach similar to that presented at the beginning of this section can be made for $v_s = 0$. Assuming again a Gaussian beam profile and no light penetration into the solid, i.e. $f(z) = \delta(z)$, a simple integration of the Green's function allows one to obtain the center temperature rise at the surface, i.e. the rise for $r = 0$ and $z = 0$. For temperature independent parameters and radial heat flow, this temperature rise can be written as

11

$$\Delta T_c = \frac{2^{1/2} \, P(1-R)}{\pi^{3/2} \, \kappa \, w_0} \; \text{arctan} \; \gamma \qquad\qquad (2.15)$$

with $\gamma \equiv 2(2D_t t)^{1/2}/w_0$ and $t \leqslant \tau$; t is the time from the beginning of the (rectangular) pulse and τ is the pulse duration. For $\gamma \ll 1$, i.e. for very short times t, (2.15) yields

$$\Delta T_c = \frac{4P(1-R)t^{1/2}}{\pi(\pi\kappa\rho \; c_p)^{1/2} \, w_0^2} \;, \qquad\qquad (2.16)$$

while for very long times (2.15) approaches the limit (2.10). The cooling cycle can be described, in the same approximation as (2.15), by

$$\Delta T_c = \frac{2^{1/2} \, P(1-R)}{\pi^{3/2} \, \kappa w_0} \; \text{arctan}\left(\frac{\gamma - \varsigma}{1 + \gamma\varsigma} \right) \qquad\qquad (2.17)$$

with $\varsigma = 2\lfloor 2D_t(t-\tau)\rfloor^{1/2}/w_0$ and $t > \tau$.

Figure 2.3 shows the evolution of the normalized center temperature for the heating (full curve) and cooling (dashed curves) cycle and for different laser pulse lengths τ. Figure 2.4 shows the radial temperature distribution for various laser beam illumination times.

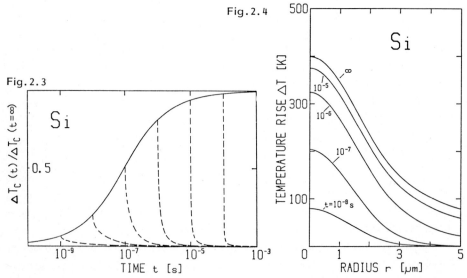

Fig.2.3. Time dependence of the increase (full curve) and decrease (dashed curves) in center temperature rise normalized to the value for continuous irradiation ($t = \infty$). $P = 0.6$ W, $2w_0 = 5$ μm, $T_0 = 300$ K, $\kappa = \kappa(Si,T_0)$, $R = R(Si,T_0)$, $D_t = D_t(Si,1680$ K) (after ⌊2.5⌋)

Fig.2.4. Distribution of temperature rise for various durations of laser light irradiation. $P = 0.6$ W, $2w_0 = 5$ μm, $T_0 = 300$ K, $\kappa = \kappa(Si,T_0)$, $R = R(Si,T_0)$, $v_s = 0$ (after ⌊2.5⌋)

12

b) Finite Absorption

Modeling the laser-induced temperature rise is a little more complicated in cases where the laser light has a finite penetration depth and where the function f(z), which enters the source term Q, must be described by f(z) = α exp(-αz) instead of by a delta function. The situation becomes even more complicated when the absorption coefficient α is not a constant but a strong function of temperature. This is the case for Si. Here, the absorption coefficient for photon energies below 3 eV, i.e. above about 410 nm, increases exponentially with temperature and can be fitted, within the range 300 K ⩽ T ⩽ 1000 K ⌊2.41-43⌋ by

$$\alpha(T) = \alpha_0 \exp(T/T_R) \ . \tag{2.18}$$

The parameters α_0 and T_R are listed in Table 2.1. Because of the drastic increase of absorption coefficent with temperature, the penetration depth α^{-1} shrinks rapidly, and this effect is even further increased by the decrease in $D_t(T)$ with increasing temperature, see (2.14).

Let us consider this in more detail for CO_2 laser and visible laser radiation. At 300 K, the infrared light of a CO_2 laser (λ ~ 10 μm, corresponding to a photon energy of about 0.12 eV) is only weakly absorbed in pure crystalline Si (c-Si; note that we frequently denote crystalline Si simply by Si, while amorphous Si is always denoted by a-Si), which has a band

Table 2.1. Parameters determining the temperature dependence of the absorption coefficent (2.18) for Si

λ [nm]	α_0 [10^3 cm^{-1}]	T_R [K]	Ref.
10000	2×10^{-5}	110	[2.42]
694	1.34	427	[2.41]
633	2.08	447	[2.41]
532	5.02	430	[2.41]
515	6.28	433	[2.41]
488	9.07	438	[2.41]
485	9.31	434	[2.41]
458	14.5	429	[2.41]
405	55.1	420	[2.41]
308	1400 ($T \leqslant 1100$ K)	4545	[2.43]
	1800 ($T > 1100$ K)		

gap of about 1.1 eV at this temperature. In undoped material the absorption coefficient can be as low as 0.3 cm^{-1} (300 K), but it can rise to more than 10^3 cm^{-1} (300 K) in heavily doped c-Si. The absorption is due to the excitation of free carriers in the conduction band. The excitation energy is transferred rapidly to the lattice via electron-phonon scattering mechanisms, which are extremely fast, typically of the order of 10^{-12} to 10^{-13} s. As a result, the lattice is locally heated, even at low to medium laser irradiances, and the absorption coefficient thereby increases according to (2.18). Simultaneously, $D_t(T)$ is decreased. This dynamic feedback due to the coupling of the optical absorption and thermal conduction rapidly increases the heating rate.

When one uses visible laser radiation, electron-hole pairs are generated. In this case, the time for energy transfer to the lattice depends strongly on carrier density, doping level, defect density, etc., and varies, typically, from 10^{-13} to 10^{-6} s. In most laser processing situations, however, recombination by Auger processes can occur, and the time of energy transfer to the lattice is again very rapid, typically of the order of some picoseconds ⌊2.44⌋. In other words, the situation is similar to that described for infrared radiation. For cw laser irradiation, a steady state between the supply and loss of energy within the irradiated region is obtained.

The effect of dynamic feedback is more pronounced for pulsed laser irradiation. A detailed theoretical treatment for Si and visible pulsed laser irradiation was devised by KWONG and KIM ⌊2.23-26⌋. The calculations use a parametrized perturbation scheme and assume one-dimensional heat flow. The rapid increase in absorption and decrease in diffusivity with pulse duration is treated in terms of effective values. The main results are summarized in Figs.2.5 and 2.6. Figure 2.5 shows the rise in center temperature for different laser pulse intensities. At 10 MW/cm^2, the effective thermal diffusion time $\tau_D = 4\alpha_o^2 D_{to} t$ (α_o and D_{to} are the effective values of the absorption coefficient and the thermal diffusivity, respectively) is long enough to increase the heated volume beyond the energy deposition depth α^{-1}. Therefore, the temperature rise near the surface is small. With increasing pulse intensity, that is with increasing heating rate, however, the reduction of D_{to} and the shrinking of α^{-1} confine the deposited energy more and more near the surface. Figure 2.6 shows the threshold laser energy for the onset of surface melting of Si as a function of pulse intensity for different laser wavelengths. At a wavelength of 694 nm and 20 MW/cm^2, for example, the energy needed to melt the Si surface is about 0.42 J/cm^2, while at 100 MW/cm^2 only

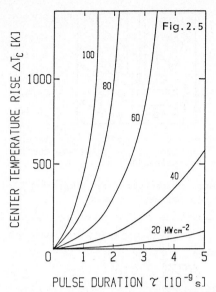
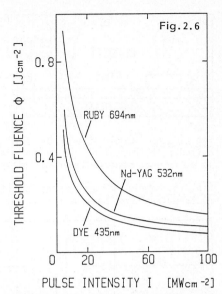

Fig.2.5. Center temperature rise for Si versus pulse duration for different pulse intensities at λ = 694 nm (after [2.23])

Fig.2.6. Threshold fluence for surface melting of Si versus pulse intensity for different laser wavelengths (after [2.23])

0.2 J/cm^2 are needed. The figure also shows the decrease in threshold pulse energy with decreasing wavelength. This effect is most pronounced in the low intensity region. It should be emphasized, however, that the calculations [2.23] did not incorporate the jump in surface reflectivity which occurs at the melting point. This may raise the threshold pulse energies for the onset of melting by a factor of approximately 2 relative to Fig.2.6.

2.1.2 Models for Deposition and Etching

As already mentioned at the beginning of this chapter, the calculation of temperature distributions for LCP is much more complicated than for the semiinfinite substrate: the temperature within the processed area will change strongly, even at constant laser irradiance, due to material parameter changes originating from the deposition, transformation or etching process itself. The essential features can be directly understood by comparing temperature distributions induced on semiinfinite substrates with those induced on structures that are relevant in LCP. Figures 2.7-9 show model structures for the deposition of circular spots, for direct writing of stripes, and for the etching of holes or grooves. For such structures,

15

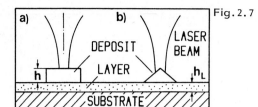 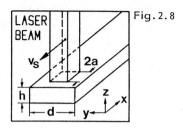

Fig.2.7a,b. Model structures for the deposition of spots. (a) circular cylinder (b) circular cone. The diameter of the spot at the substrate surface is d. An intermediate layer of thickness h_L is indicated

Fig.2.8. Model structure for steady growth of stripes. For simplicity, a rectangular laser beam with constant intensity has been assumed

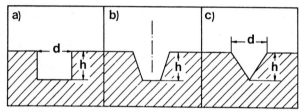

Fig.2.9a-c. Model structures for the etching of holes. Here, d is defined as the width at the substrate surface

analytic solutions of the heat equation (2.1) are, in general, not possible. Therefore, numerical methods such as finite difference or finite element procedures must be employed [2.45,46]. The first calculations of this kind were performed by PIGLMAYER et al. [2.1-5]. The boundary conditions were similar to those used in the preceding section. The approximation was made that at sufficiently large distances from the irradiated surface, the temperature rise ΔT becomes radially symmetric with respect to the center of the laser spot.

Let us commence by simulating the deposition of circular *spots*. Here, we do not consider the problems in the phase of nucleation, as discussed in Sect.5.1, but assume that after a time t_m the semiinfinite plane substrate is already covered with a thin circular film of the deposited material within the area exposed to the focused laser beam. In Fig.2.7a, the deposit is represented by a circular disc of diameter d and height h. For greater generality, the semiinfinite substrate is assumed to be covered with a thin extended layer of thickness h_L. The thermal conductivities of the deposit, the thin layer and the substrate are κ_D, κ_L and κ_S. For simplicity we ignore the temperature dependence of these material parameters, and also heat losses to the gas phase and latent heat effects (heats of formation). The laser beam is assumed to be Gaussian and normally incident at the center of the deposit.

16

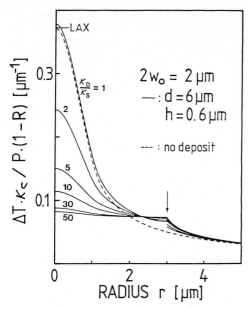

Fig.2.10. Laser-induced temperature distributions for a circular cylinder (Fig.2.7a) calculated for different ratios of thermal conductivities κ_D and κ_S. The arrow marks the edge of the disc. The dashed curve represents the temperature distribution for a semiinfinite substrate. The center temperature according to analytical calculations by LAX [2.7] is also marked (after [2.2])

First, we make the following further assumptions: the diameter of the disc is larger than the laser focus ($d > 2w_0$); no light penetrates into the deposit [$f(z) = \delta(z)$]; the thermal conductivities of the thin layer and the substrate are equal ($\kappa_L = \kappa_S$). Under these conditions, the temperature distribution can be calculated from (2.4,5) together with (2.6') and $v_s = 0$. The only difference from the semiinfinite substrate appears in the source term Q. Here, the reflectivity of the substrate must be replaced by the reflectivity of the deposited material, and the thermal conductivity now refers to both the substrate *and* the deposit, i.e. κ assumes the values κ_D and κ_S in the deposit and the substrate, respectively. Figure 2.10 shows the results of numerical calculations for various ratios of the thermal conductivities κ_D and κ_S ($d = 6$ μm, $h = 0.6$ μm, $2w_0 = 2$ μm). The edge of the deposit is indicated in the figure by the arrow. It can be seen from the figure that the center temperature scales approximately with κ_D/κ_S. This holds also for $\kappa_D < \kappa_S$, a case that under certain circumstances applies to Si deposition on Si substrates [see (2.11)]. For $\kappa_D = \kappa_S$, the temperature distribution within the deposit is close to that for the plane substrate (dashed curve); the center temperature taken from the analytical calculations by LAX [2.7] is indicated. Significant differences from the semiinfinite substrate occur only near the edge of the disc. With increasing κ_D/κ_S the temperature distribution on the disc flattens. In the limit $\kappa_D \gg \kappa_S$ the temperature is almost constant over the disc. This latter case applies, for

17

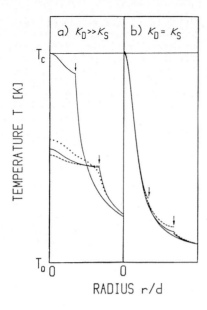

Fig.2.11a,b. Temperature distributions calculated for different geometries of circular deposits. The dotted curve refers to a circular cone (Fig.2.7b), all other curves to circular cylinders (Fig.2.7a). Arrows indicate edges of deposits. Full curves: h = d/20, $2w_0$ = d/3 (for the upper full curves the values for h and d are half as large as for the lower ones). Dashed curve: h = d/10, $2w_0$ = d/3. Dotted curve: h = d/20, $2w_0$ = d/3. (a) κ_D = 70 W/mK, κ_S = 1.3 W/mK; (b) κ_D = κ_S = 70 W/mK. T_0 = 300 K (after [2.1])

example, to metal deposits on insulating substrates. The temperature distribution outside of the deposit, i.e. for r > d/2, is only slightly influenced by the value of the ratio κ_D/κ_S.

Next, we consider temperature distributions for different geometries of circular deposits. They are shown in Figs.2.11a,b for two different ratios of thermal conductivities κ_D and κ_S. The value for κ_D corresponds to a typical metal such as Ni [κ_D(Ni) ≈ 70 W/mK]. The substrates are glass [κ_S(glass) ≈ 1.3 W/mK] and Si [κ_S(Si) ≈ κ_D(Ni)] in Figs.2.11a and b, respectively. All curves except the dotted one refer to the model of the circular cylinder (Fig.2.7a). The dotted curve is the temperature profile for a circular cone (Fig.2.7b) with height ḣ = d/20 (in Fig.2.11b the curves would be almost indistinguishable). Figure 2.11a shows that for constant spot diameter d, the temperature at the edge of the deposit, T(d/2), depends only very slightly on its geometry, i.e. on its height and exact shape (compare full, broken and dotted curves). Actually, if κ_D >> κ_S, the temperature rise ΔT(d/2) scales approximately inversely with d (compare full curves) and can be described to a good approximation by the simple equation [2.29]

$$\Delta T(d/2) = P\ (1-R_D)/(2d\kappa_S)\ . \tag{2.19}$$

It is therefore not surprising that ΔT(d/2) is not very sensitive to changes in κ_D as long as κ_D >> κ_S; e.g., when we use a value of κ_D = 30 W/mK instead of κ_D = 70 W/mK, ΔT(d/2) decreases by only about 5%. In Fig.2.11b the temperature distribution is nearly independent of the diameter, height and shape of the deposit and is very similar to that for a plane substrate; significant differences occur only near the edge of the deposit.

Figure 2.12 shows the influence of the laser focus on the temperature distribution for three different ratios of thermal conductivities κ_D, κ_L and κ_S. The values used for κ_D and κ_S are the same as those used in Figs.2.11a,b. Figure 2.12b represents an intermediate case where the Si substrate is covered with a layer of h_L = 4000 Å a-SiO$_2$ $\lfloor \kappa_L(SiO_2) \approx \kappa(glass) \rfloor$. The full curves were calculated for equal center temperature T_C = 530 K, which of course requires different absorbed laser powers, namely 10, 56 and 120 mW in cases a, b and c, respectively. A doubling (dash-dotted curves) of the laser focus changes the center temperature much more dramatically in b and c than in a. As long as d > 2w_0, the temperature $\Delta T(d/2)$ at the edge of the deposit remains nearly unaffected in all three cases.

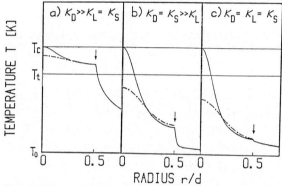

Fig.2.12a-c. Temperature distributions calculated for model structure shown in Fig.2.7a with three different ratios of thermal conductivities and two different radii of the laser focus. Full curves have been calculated for equal center temperature with h = d/20, 2w_0 = d/3. Dash-dotted curves: h = d/20, 2w_0 = 2d/3. Values for κ_D and κ_S correspond to those in Fig.2.11. h_L = 4000 Å, T_0 = 300 K, T_C = 530 K. T_t schematically indicates a threshold temperature (after $\lfloor 2.1 \rfloor$)

We now proceed to a model for *direct writing* (see Chap.5). The shape of the stripe, shown in Fig.2.8, has been chosen for mathematical convenience; for the same reason, the intensity of the incident beam is assumed in this case to be constant over its square cross section. Relative to the scanning beam the shape of the infinitely long stripe is taken to be static; in a reference frame where the laser beam and thereby the stripe are at rest the substrate moves with velocity v_s. Temperature profiles calculated for κ_D = 30κ_S and κ_D = κ_S are shown in Fig.2.13. For κ_D = κ_S, the temperature

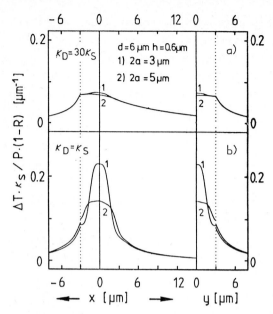

Fig.2.13. Calculated temperature distributions for stripes along the coordinate axes of the model structure shown in Fig.2.8. v_s/D_t < 0.5 μm^{-1} (after $\lfloor 2.2 \rfloor$)

distribution is again not significantly affected by the deposit and is almost symmetric. For $\kappa_D > \kappa_S$, the general trend is the same as for the discs. For realistic geometrical parameters for the width d and the thickness h of stripes, and for v_s/D_t small compared to a typical parameter v_α, the temperature profiles are essentially unaffected by the velocity of the laser beam. For example, for parameters d = 6 μm, h = 0.6 μm we obtain $v_\alpha \approx 0.5$ μm^{-1}, which corresponds to scanning velocities of 50 m/s on Si and 0.50 m/s on glass. These values exceed realistic scanning velocities in laser-induced deposition by several orders of magnitude. The temperature profiles shown in Fig.2.13 are therefore very similar to those shown in Fig.2.12. The main differences in Fig.2.13 result from the heat transport along the stripe (positive x-direction), which yields a reduction of the center temperature with increasing cross section of the stripe. This effect is especially significant for $\kappa_D \gg \kappa_S$ (Fig.2.14). The temperature distributions shown in Figs.2.10-14 will be further discussed in Chap.5.

For the model structures shown in Figs.2.9a-c, which are relevant in laser-induced *etching*, only preliminary calculations have been performed $\lfloor 2.5,6 \rfloor$. The case of Fig.2.9a is very simple as long as d > $2w_o$. Under this condition, the temperature distribution is similar to that of a semiinfinite

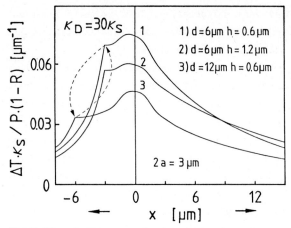

Fig.2.14. Influence of the cross section of stripes on the temperature profile. The meaning of the dashed arrows is explained in Sect.5.2.5 (after [2.2])

substrate. The situation becomes much more difficult when the width d becomes comparable to the laser focus $2w_0$. For the case of the V-shaped hole shown in Fig.2.9c, the calculations are more complicated even for $d > 2w_0$. If we neglect multiple reflections of the laser light within the hole, we obtain a temperature distribution that shows a marked dip at the center of the hole, i.e. for $r = 0$. This would imply, of course, that further laser beam illumination would change the shape of the hole. This contradicts the experimental results in Chap.8. To maintain the maximum temperature rise at the center, one has to take into account the dependence of the reflection and absorption coefficients on the angle of incidence and, additionally, multiple reflections of the laser light inside the hole. A comparison with the experimental results in Chap.8 requires consideration of another point. While the depths h of holes increase continuously with laser beam illumination time t_i, the widths d remain about constant. Consequently, the angle of incidence of both the incoming laser light and the internally reflected laser light will also change with time. Similar difficulties arise for the model in Fig.2.9a if d becomes $< 2w_0$, and also for the intermediate case shown in Fig.2.9b. Therefore, the distribution of the absorbed laser power within the hole becomes very complicated and differs significantly from the intensity distribution within the laser beam. In most cases, the real situation in laser-induced etching is even more complex. The material that is ejected out of the hole during the etching process scatters and/or partially absorbs the incoming and internally reflected laser light. These difficulties may explain why calculated temperature distributions for the modeling of laser-induced etching have not so far proved very enlightening. This is quite different from the case of laser-induced deposition, where many experimental results can be understood from the calculations outlined above.

The accuracy of the model calculations is closely related to the knowledge of the parameters R, κ, α, etc. that enter these calculations. These parameters often depend on the temperature itself; one must then perform calculations self-consistently. Additionally, these parameters may change with the laser wavelength, the microstructure and morphology of the material within the processed area, the film thickness, impurities, etc. It is therefore desirable to measure these quantities, or at least some of them, in situ, i.e. during the deposition or etching process. For a few model systems such measurements have in fact been performed for R and α. In some cases, additional estimations can be made. For example, the reflectivity of rough surfaces does not depend upon the depth or the spatial period of the roughness independently, but rather upon their ratio. This has been studied by KIVAISI and STENSLAND [2.47]. By comparing morphologies of deposits derived from scanning electron micrographs with the results in [2.47], additional information on the value of R may be obtained.

2.2 Photolytic Processing

In photolytic (photochemical) processing, the laser light breaks chemical bonds directly within the surface of the material itself, within adspecies or within the surrounding gaseous or liquid medium. Photochemical bond-breaking can be based on dissociative electronic excitations, which are located in the visible and ultraviolet spectral region, or on *selective* multiphoton vibrational excitations by means of infrared radiation. An indirect mechanism is to transfer the bond-breaking energy via an intermediate species as, for example, in photosensitization. Clearly, single- or multiphoton dissociation of molecules based on direct or indirect electronic excitation is a nonthermal and very common process. Photochemical dissociation of molecules based on selective multiphoton vibrational excitations is not as common and in fact very rare in LCP. The fine details of the different fundamental mechanisms have been extensively studied for model systems and frequently discussed in a great variety of monographs and conference proceedings (see e.g. [2.48-61]). On the other hand, apart from a very few exceptions, only little is known about the photochemistry of those molecules that are relevant in LCP. Additionally, in most LCP situations, the physical conditions with respect to molecular densities, temperatures, the great variety of species, the presence of interfaces or acceptors, etc., differ significantly from those generally chosen for separating and investigating single interaction mechanisms. Selective bond breaking in the condensed phase is not yet well understood and difficult to realize because of line broadenings and fast vibrational relaxations into heat, typically within 10^{-13} - 10^{-11} s. In the following outline we will confine ourselves to some fundamentals, with special emphasis on the literature on those gas-phase molecules that are used as precursors in LCP. Solid-phase excitation will be discussed further in Chap.3.

2.2.1 Dissociative Electronic Excitations

Electronic dissociation of molecules can be based on single- or multiphoton processes. Such processes are, in general, accompanied by simultaneous excitations of vibrational and rotational transitions (see e.g. [2.48-64]).

Let us start with four characteristic cases of dissociative single-photon excitations. Figure 2.15 shows schematically potential energy curves for the electronic ground state and excited states of different molecules. For simplicity, vibrational and rotational energy levels are not included. According to the Franck-Condon principle, transitions always occur vertically between maxima in $|\Psi_1|^2$ and $|\Psi_2|^2$, where the Ψ_i are the corresponding vibrational wave functions in the lower and upper electronic states. In the

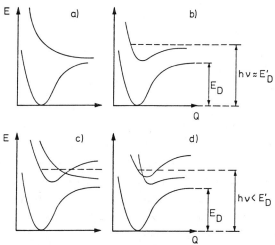

Fig.2.15a-d. Potential energy curves for the electronic ground state and excited states showing different cases of dissociation. Here, E_D is the energy of dissociation and $h\nu$ the photon energy. Vibrational and rotational energy levels are not included in the curves

case of Fig.2.15a the excited electronic state is unstable and the molecule, after having been brought into the unstable excited state, typically dissociates within 10^{-13} - 10^{-14} s. Clearly, relaxation and energy transfer between gas-phase molecules is unlikely within such a short time. In the case of Fig.2.15b the excited electronic state is stable and dissociation only occurs for photon energies $h\nu > E_D'$. However, in many cases dissociation is even observed for $h\nu < E_D'$ (Figs.2.15c,d). This phenomenon is called spontaneous predissociation. Here, dissociation occurs before the dissociation limit of the initially excited electronic state by transitions to another unstable (Fig.2.15c) or less stable (Fig.2.15d) electronic state. Another mechanism is dissociation by internal conversion into vibrational states of the electronic ground state having an energy above E_D. Such transitions become possible if a mixing of states near crossings of potential curves occurs. Predissociation is therefore much more frequent in polyatomic molecules than in diatomic molecules. The typical time scale for predissociaton extends from 10^{-6} to 10^{-12} s. For low light intensities the average number of dissociated molecules N_d is proportional to the laser fluence ϕ and is given in good approximation by $N_d = \sigma_d^{eff} N \phi/h\nu$, where N is the total number of molecules within the interaction volume and $\phi = I\tau$, I being the laser light intensity and τ the laser beam illumination time. The effective cross section for dissociation is $\sigma_d^{eff} = \eta\sigma_d$. The dissociation yield η depends on the gas pressure, the type of reactant and buffer gas molecules, etc. Under collisionless conditions, $\eta = 1$. For $h\nu > E_D'$ and negligible fluorescence, the dissociation and absorption cross sections are equal, i.e. $\sigma_d \approx \sigma_a$. In the case of linear interaction, this number of excited or dissociated molecules is independent of whether irradiation is pulsed or continuous.

The main limitation of single-photon decomposition processes relevant to laser-induced chemical processing is the lack of flexibility of available lasers in the medium to far ultraviolet spectral region. Multiphoton processes impose less severe wavelength restrictions on the absorption continuum of the reactive species than single-photon processes. However, in this case the number of excited molecules depends nonlinearly on the photon flux, and hence on the excitation conditions. Efficient processing can only be performed with high power pulsed lasers. Here, other problems may arise in many processing applications (Sect.4.2).

In the following, we will become more specific and concentrate on the *homogeneous* photochemistry of molecules that are most commonly used in deposition, surface modification and etching reactions. A brief glance at

Table 2.2. Dissociation (d) and absorption (a) cross sections σ_d^{eff} and σ_a for parent molecules used in LCP

Molecule	σ [$\times 10^{-18}$ cm^2]	λ [nm]	Ref.
Al$_2$(CH$_3$)$_6$	20 (a)	193	[2.75]
	0.002 (a)	257	[2.75]
AsH$_3$	18 (a)	193	[2.66]
B$_2$H$_6$	0.2 (a)	193	[2.66]
Cd(CH$_3$)$_2$	2 (a)	257	[2.75]
Cr(CO)$_6$	12 (a)	193	[2.72]
	33 (a)	249	[2.72]
	5.2 (a)	308	[2.72]
Fe(CO)$_5$	240 (a)	193	[2.67]
	27 (a)	248	[2.67]
	1.3 (a)	355	[2.67]
Ga(CH$_3$)$_3$	5.4 (a)	193	[2.68]
	0.09 (a)	257	[2.68]
GeH$_4$	0.0035 (a)	193	[2.32]
InI	<7 (d)	193	[2.69]
Mo(CO)$_6$	60 (a)	193	[2.72]
	44 (a)	249	[2.72]
	11 (a)	308	[2.72]
	0.5 (a)	350 – 360	[2.70]
NF$_3$	0.0053 (a)	193	[2.71]
Ni(CO)$_4$	30 (a)	248	[2.136]
	2.4 (a)	308	
PH$_3$	13 (a)	193	[2.66]
Pt(PF$_3$)$_4$	0.19 (a)	248	[2.73]
Pt(HFAcAc)^a	100 (a)		[2.74]
SiH$_4$	0.0012 (a)	193	[2.66]
TlBr	22 (d)	193	[2.69]
TlI	24 (d)	193	[2.69]
	2.6 (d)	248	[2.69]
W(CO)$_6$	12 (a)	193	[2.72]
	4.5 (a)	249	[2.72]
	2.4 (a)	308	[2.72]
	0.5 (a)	350 – 360	[2.70]

^{a} HFAcAc is CF$_3$COCHCOCF$_3$ i.e. the 1,1,1,5,5,5 hexafluoro-2,4-pentane-di-onate anion, which is also known as the hexafluoroacetylacetonate anion.

Tables 5.1, 6.2 and 8.1 reveals that the most important precursor molecules involve alkyls, carbonyls, halides and hydrides. Table 2.2 summarizes dissociation (d) and absorption (a) cross sections for certain compounds at specific laser wavelengths. It becomes evident that the values of σ may differ, at a certain wavelength, by several orders of magnitude. For example, for $Cd(CH_3)_2$ and $Al_2(CH_3)_6$ (the molecule is dimerized at 300 K and pressures of several millibars) the absorption cross sections at λ = 257 nm differ by a factor of about 10^3. This can be derived from the UV absorption spectra presented in $\lfloor 2.75 \rfloor$. Efficient dissociation of $Al_2(CH_3)_6$ requires irradiation at a shorter laser wavelength. It is absolutely essential to note that the dissociation yield depends strongly on gas pressure and on possible admixtures. Furthermore, LCP is often performed not homogeneously, but at gas-solid or liquid-solid interfaces. Therefore, laser-molecule-surface interactions are of great importance. As a consequence, the photodissociation yield may change by orders of magnitude with respect to collisionless unimolecular reactions or homogeneous collision-induced reactions within the gas or liquid phase. Various light-molecule-surface interaction mechanisms are discussed in Chap.3.

Metal alkyls, such as $Al_2(CH_3)_6$, $Cd(CH_3)_2$, $Zn(CH_3)_2$, have been used for the deposition of the corresponding metals (Sects.5.2 and 5.3) and, additionally, as doping gases (Sect.6.3). Admixtures of these molecules with N_2O or NO_2 allow formation of oxide layers (Sects.5.3.2 and 5.3.3). The molecules $Ga(CH_3)_3$, $P(CH_3)_3$, $Zn(CH_3)_2$, $(CH_3)_3InP(CH_3)_3$, etc. are precursors for the deposition of compound semiconductors (Sect.5.3.2), and many of them show dissociative continua in the near to medium ultraviolet, which can easily be reached with available laser sources. The photothermal and photochemical decomposition kinetics of many metal alkyls and carbonyls has been reviewed by PRICE $\lfloor 2.64 \rfloor$. For most of the metal alkyls, however, little is known about the details of the various photofragmentation channels. An exception is $Cd(CH_3)_2$, which has been studied both theoretically $\lfloor 2.76 \rfloor$ and experimentally $\lfloor 2.77 \rfloor$. According to these investigations, single-photon absorption near 257 nm (frequency-doubled Ar^+ laser radiation) results in dissociation into ground state Cd and CH_3 according to

$$Cd(CH_3)_2 + h\nu(257 \text{ nm}) \longrightarrow Cd(^1S_0) + 2CH_3 \ . \tag{2.20}$$

Dissociation is asymmetric, with one CH_3 group leaving at least half a period of vibration later than the first. The free methyl radicals subsequently react to form volatile hydrocarbons such as ethane.

Metal carbonyls, such as $Ni(CO)_4$, $Fe(CO)_5$, $Cr(CO)_6$, $Mo(CO)_6$, and $W(CO)_6$, were used for metal deposition in the form of both microstructures and extended thin films. The photodissociation dynamics of these molecules has been studied in some detail ⌊2.64,67,78-84⌋. For many metal carbonyls, molecular fragmentation begins to occur in the near UV region at wavelengths $\lambda < 350$ nm. For cw and pulsed laser irradiation at low power densities, decomposition seems to be based on *sequential* elimination of CO ligands by single photon processes, such as

$$Me(CO)_m + h\nu \longrightarrow Me(CO)^*_{m-1} + CO , \qquad (2.21a)$$

$$Me(CO)^*_{m-1} + h\nu \longrightarrow Me(CO)^*_{m-2} + CO , \qquad (2.21b)$$

$$\qquad \bullet \qquad \bullet \qquad \bullet \qquad \bullet$$
$$\qquad \bullet \qquad \bullet \qquad \bullet \qquad \bullet$$
$$\qquad \bullet \qquad \bullet \qquad \bullet \qquad \bullet$$

$$Me(CO)^* + h\nu \longrightarrow Me^* + CO , \qquad (2.22)$$

where * indicates internal vibrational and possibly electronic excitation. Because of the high gas pressures used in LCP, stripping of the remaining ligands, for example after absorption of two photons according to (2.21a) and (2.21b), can also occur by subsequent collisional events. YARDLEY et al. ⌊2.67⌋ have investigated the photolysis of $Fe(CO)_5$ at 352 nm (TH Nd:YAG), 218 nm(KrF) and 193 nm (ArF). At low laser fluences and gas pressures below 130 mbar, the primary fragmentation into $Fe(CO)_{5-n}$ with $1 \leqslant n \leqslant 4$ proceeded via one-photon absorption. A large fraction of the initial energy was found to have been retained as internal electronic and vibrational energy in the metal-containing fragment. The photofragmentation of $Cr(CO)_6$ has been studied recently by SEDER et al. ⌊2.79⌋. Under 351 nm XeF laser radiation the major photolysis product was $Cr(CO)_5$. The large relative yield of $Cr(CO)_5$ is consistent with the observations of BRECKENRIDGE and SINAI ⌊2.80⌋ who used 355 nm frequency-tripled pulsed Nd:YAG laser radiation. Focused high power pulsed laser excitation may favor *coherent* multiphoton rather than sequential single-photon photochemistry. FISANICK et al. ⌊2.83⌋ have studied the multiphoton dissociation (MPD) and ionization (MPI) of $Cr(CO)_6$ and related compounds. Different fragmentation channels have been observed for parent molecule excitation below and above the ionization limit corresponding to a wavelength of about 400 nm.

The photochemistry of organometallic coordination complexes has recently become the object of great interest. Such compounds have been successfully

27

used as precursors for noble metal deposition. Initial investigations on the generation of Cu atoms from the photodissociation of $Cu(HFAcAc)_2$ have been performed by MARINERO and JONES ⌊2.85⌋. Transition metal complexes such as CrO_2Cl_2, OsO_4 and $Pt(PF_3)_4$ have been investigated by SCHRÖDER et al. ⌊2.73,86⌋.

Halides like Cl_2, Br_2, I_2, etc. show strong continua in the visible and ultraviolet region. They result at least in part from the allowed dissociative transition $^1\Pi_u \longleftarrow {}^1\Sigma_g$, for example

$$Cl_2 + h\nu \ (\lesssim 500 \text{ nm}) \longrightarrow 2Cl \ . \tag{2.23}$$

For Cl_2, the maximum in the dissociative continuum occurs at about 330 nm. At wavelengths $\lambda > 480$ nm, the continuum is very weak and has vibrational structure superimposed on it, resulting from transitions into the bound $^3\Pi(0_u^+)$ state. If absorption occurs at a wavelength short enough to break the bond ($\lambda < 498.9$ nm), this bound state predissociates with near unity yield by crossing over to a repulsive state ⌊2.87,88⌋. For Br_2 and I_2, wavelengths below 628.4 nm and 803.7 nm are necessary to dissociate or predissociate the molecule. The halogen radicals are very aggressive. They strongly chemisorb on many surfaces and may thereby break surface chemical bonds. Photoreactions such as (2.23) are therefore often used in materials etching (Chap.8). Similarly, photochemical etching is also often performed with halogen compounds such as COF_2, CF_2Cl_2, CF_2Br_2, CF_3Br, CF_3I, CF_3NO, $CO(CF_3)_2$. These molecules can be photodissociated with ArF or KrF excimer laser radiation resulting in highly reactive radicals such as F, Cl, CF_2, CF_3 ⌊2.48,52,57,89,90-92⌋. Ultraviolet absorption spectra of many of these compounds are shown in ⌊2.90-92⌋. The pyrolytic and photolytic decomposition kinetics of many halides has been reviewed by ARMSTRONG and HOLMES [2.93⌋. Admixtures of metal halides and hydrogen are used for metal deposition. The most important precursor molecule so far is WF_6. This molecule is most commonly used for photodeposition of extended thin films of W, mainly by means of ArF excimer laser radiation. DEUTSCH and RATHMAN ⌊2.94,95⌋ suggest that the ArF laser radiation initially produces WF_n (n = 1-5) radicals which further react with hydrogen (Sect.5.3.1).

Hydrides such as SiH_4, Si_2H_6, GeH_4, CH_4, C_2H_2 are used for deposition of Si, Ge and C. Photolysis of admixtures of SiH_4 with N_2O is important in thin film formation of SiO_2 (Sect.5.3.3). AsH_3, B_2H_6, and PH_3 are used as precursors for silicon doping. Direct photochemical decomposition of silicon hydrides can occur only for wavelengths of less than 200 nm. PERKINS et al. ⌊2.96⌋ have studied the 147 nm photolysis of SiH_4, using a Xe resonance lamp.

Two primary decomposition processes have been revealed

$$SiH_4 + h\nu \longrightarrow SiH_2 + 2H \qquad\qquad (2.24)$$

and

$$SiH_4 + h\nu \longrightarrow SiH_3 + H . \qquad\qquad (2.25)$$

The final products of the photodecomposition reaction were H_2, Si_2H_6, Si_3H_8 and a film of hydrogenated amorphous silicon (a-Si:H; see also Sect.5.3.2). The shortest laser wavelength that has been used for the decomposition of SiH_4 is 193 nm, the ArF excimer laser line; no experiments with 157 nm F_2 laser radiation are known (Table 4.1). At 193 nm, SiH_4 has negligible linear absorption. However, the molecule can be dissociated at this wavelength with relatively low fluences (< 10 MW/cm^2) by multiphoton excitation. ArF laser photolysis of C_2H_2 has been investigated by IRION and KOMPA [2.97].

At wavelengths less than 200 nm, the primary photolysis steps for AsH_3, B_2H_6 and PH_3 are thought to include production of AsH_2, B_2H_5, PH_2 and atomic H [2.66]. In the presence of SiH_4, the H atoms may react with SiH_4 according to

$$SiH_4 + H \longrightarrow SiH_3 + H_2 . \qquad\qquad (2.26)$$

The photochemistry of N_2O and NH_3 has met with increasing interest in connection with the deposition of extended thin films of oxides and nitrides (Sect.5.3.3). For N_2O the quantum yield for dissociation is about 1 in the wavelength region 138 - 210 nm [2.98]. Its photochemistry has been well categorized for single-photon excitation [2.98-103]. The primary reactive product under ArF excimer laser irradiation [2.103] is excited atomic oxygen

$$N_2O + h\nu(193 \text{ nm}) \longrightarrow N_2 + O(^1D) . \qquad\qquad (2.27)$$

The photodissociation of NH_3 under 193 nm ArF laser irradiation has been investigated by DONNELLY et al. [2.104]. The primary photoproduct is ground state NH_2, which is formed with nearly unit efficiency.

2.2.2 Photosensitization

In photosensitization the photons are directly absorbed by intermediate species which then deexcite by collisional transfer of the appropriate energy to the acceptor molecules (see e.g. [2.57]). For example, direct photolysis of CH_4 is only possible below 144 nm

$$CH_4 + h\nu(<144 \text{ nm}) \longrightarrow CH_2 + H_2 \, , \tag{2.28}$$

while the Hg-photosensitized reaction can take place at a longer wavelength

$$Hg(^1S_0) + h\nu(253.7 \text{ nm}) \longrightarrow Hg(^3P_1) \, , \tag{2.29}$$

$$Hg(^3P_1) + CH_4 \longrightarrow Hg(^1S_0) + CH_3 + H \, . \tag{2.30}$$

As can be seen from this example, the products of these photoreactions may differ greatly in both cases. Mercury-photosensitized decomposition of SiH_4 and GeH_4 has been studied by NIKI and MAINS ⌊2.105⌋ and by ROUSSEAU and MAINS ⌊2.106⌋, respectively. Photosensitized reactions are very common in photochemical studies, but are not favored for localized microchemistry due to the reaction-spreading properties intrinsic to the process (see also Chap.9). However, the technique can be used for large-area processing. Examples are the low temperature growth of epitaxial layers of HgTe ⌊2.107,108⌋ and etching reactions with CF_3, produced from $CO(CF_3)_2$ via photosensitization of C_6H_5F by KrF excimer laser light ⌊2.90-92⌋.

2.2.3 Infrared Vibrational Excitations

In this section we will briefly outline some fundamentals of laser-induced vibrational excitations of free molecules in the electronic ground state. Several reviews have appeared in this field in the last few years ⌊2.48-52, 58-61,109-111⌋. Here, we again put special emphasis on molecules and aspects that are relevant to LCP.

Let us commence by defining terms: We will henceforth call gas- or liquid-phase reactions thermal if the absorbed laser energy is at least locally thermalized between the different degrees of freedom. On the other hand, we will call reactions nonthermal if there are molecules participating in them that are not in local thermal equilibrium. We will use the term nonthermal or photochemical reaction even in cases where the laser light induces a local temperature rise but without complete thermalization, e.g. between vibrational and translational degrees of freedom.

Classification can be performed according to the relaxation times τ_{v-v}^{intra}, τ_{v-v}^{inter}, τ_{v-T}, and the rate of vibrational excitation W_e. The first quantity, τ_{v-v}^{intra}, is the time for intramolecular transfer of vibrational energy between different vibrational modes of the molecule being excited. This time increases with decreasing vibrational anharmonicity and is typically of the order of 10^{-12} to 10^{-11} s. The second, τ_{v-v}^{inter}, is the time for intermolecular transfer of vibrational energy between molecules of the same kind or of different kinds within a gas or a liquid. Finally, τ_{V-T} is the relaxation time for molecular vibrational energy to be transferred to translational degrees of freedom - which is the time for thermal equilibrium to be reached in the molecular mixture. Clearly, the times τ_{V-V}^{inter} and τ_{V-T} vary with experimental conditions such as the molecular density, temperature and type of admixtures or solvents. For gases such as NH_3 or BCl_3 at 1000 mbar, τ_{V-V}^{inter} is typically of the order of 10^{-9} s. The value of W_e will depend on the radiation intensity and the cross section of the specific vibrational transition. For further details see, for example, the monographs by BEN-SHAUL et al. ⌊2.48⌋ and LETOKHOV ⌊2.52⌋.

Mode- or *bond*-selective multiphoton excitation requires an excitation rate that is large compared to the rate of intramolecular vibrational energy transfer. This would need laser pumping in a mode fairly isolated from the other vibrational modes and high laser intensity picosecond or subpicosecond excitation. While the first condition is well fulfilled for diatomic molecules, which have only one vibrational degree of freedom, collisionless multiphoton dissociation by monochromatic infrared radiation seems to be impossible in this case because of the energy mismatch between the photon energy and the vibrational energy levels, which, because of the anharmonicity in the potential, are not equally spaced. This latter restriction is relaxed in the case of polyatomic molecules (see below). Nevertheless, bond-selective dissociation has not yet been convincingly demonstrated even in a collisionless environment.

In the following we consider three different cases that can be realized under conditions used in LCP. First, if

$$1/\tau_{V-V}^{intra} \gg W_e \gg 1/\tau_{V-V}^{inter} , \qquad (2.31)$$

molecule-selective excitation is possible. While the vibrational energy within the molecule interacting with the infrared field is in equilibrium, there is no vibrational equilibrium among the molecules in the mixture. In other words, molecules in resonance with the laser frequency acquire a higher vibrational temperature than all the other molecules. Molecule-selective excitation and dissociation according to (2.31) is of considerable practical interest, e.g. in laser isotope separation, and it has been studied in great detail. In order to separate single effects, many of the fundamental investigations have been performed in the collisionless environment of a molecular beam. Under this condition, τ_{V-V}^{inter} approaches infinity. For this case, we will briefly describe the dissociation process. In contrast to

31

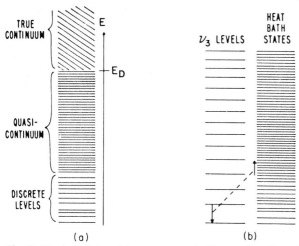

TRUE
CONTINUUM

E

E_D

QUASI-
CONTINUUM

DISCRETE
LEVELS

(a)

ν_3 LEVELS

HEAT
BATH
STATES

(b)

Fig.2.16a,b. Schematic representation of vibrational energy levels in a polyatomic molecule. (a) shows three regimes of energy levels. (b) shows the levels for the selectively driven mode ν_3, and the ensemble of other vibrational levels. Energy transfer and relaxation are caused by anharmonic coupling of modes

diatomic molecules, polyatomic molecules can absorb a great number of monochromatic photons ⌊2.48-52,59-61,109-114⌋. This can be made plausible from the schematic of vibrational energy levels shown in Fig.2.16. At very low energies the energy levels are discrete. With increasing vibrational energy and number of atoms of the polyatomic molecule, the complexity of the energy levels increases very rapidly. The energy states become very dense. This region is called the quasi-continuum, which merges into a true continuum above the dissociation limit. The quasi-continuum typically starts at the level of 1 vibrational quantum (for complex molecules and molecules with heavy atoms) to 3-10 vibrational quanta (for simple polyatomic molecules). It can be described as a heat bath of states formed by the other vibrational degrees of freedom. The dissociation process involves an initial, selective excitation of the infrared driven mode into the quasi-continuum region, followed by energy absorption by the quasi-continuum until the lowest dissociation channel is reached. Collisionless infrared multiphoton excitation and dissociation of many molecules, including SF_6, BCl_3, $CO(CF_3)_2$, CF_3I and CDF_3, which are also used in LCP, is consistent with this model. Clearly, this model describes the situation in gas mixtures equally well, as long as condition (2.31) is fulfilled. The average number of IR photons absorbed by a molecule is $<n> = \sigma\phi/h\nu$, where ϕ is the laser fluence and σ the absorption cross section. In general, σ is not a constant but itself changes with laser fluence and within the time τ of the laser pulse. The value of ϕ for which $\langle n \rangle \gg 1$ ranges from 10^{-3} J/cm^2 for complex molecules to tens of J/cm^2 for simpler molecules with large rotational constants [2.52].

Second, we consider the condition

$$1/\tau^{inter}_{v-v} \gg W_e \gg 1/\tau_{v-T} \ . \tag{2.32}$$

Here, the vibrational equilibrium among all the molecules in the mixture is stronger, but the system is still not in thermal equilibrium. Condition (2.32) can only be fulfilled if the gas mixture does not contain any component with fast v-T relaxation. Because of the difference between vibrational and translational temperature, nonselective vibrational photochemistry is possible when reactions with a minimum energy barrier take place in a time not greater than about τ_{v-T}.

The last condition is characterized by

$$W_e \ll 1/\tau_{v-T} \ . \tag{2.33}$$

In this case, all the molecules within the reaction volume defined by the laser beam are in thermal equilibrium. The vibrational energy is immediately thermalized through bi- or multimolecular collisional channels. The reaction is purely thermal. Nevertheless, laser-induced gas- or liquid-phase heating may significantly differ from traditional heating, e.g. via thermalization at or near the surface of a heated substrate. This has already been mentioned at the beginning of Sect.2.1.

After these very general remarks, we will become more specific and discuss some reactions that are of special importance in LCP. First of all, we should be aware that for the complex molecules and molecular mixtures commonly used in LCP, very little is known about the various $v \longrightarrow v$ and $v \longrightarrow T$ relaxation pathways. However, because of the complexity and variety of the molecules involved and also on account of the high molecular densities used, we can speculate that condition (2.32), and to an even greater extent condition (2.33), will apply in most cases. An important example in LCP that seems to belong to this case is the deposition of hydrogenated amorphous Si from SiH$_4$ by means of CO$_2$ laser radiation incident parallel to the substrate (see Fig.4.3 and Sect.5.3). It has been shown by MEUNIER et al. [2.115] that the deposition rate follows an Arrhenius type behavior where the temperature corresponds to the gas temperature T_g induced by *single*-photon vibrational absorption of SiH$_4$ and collisional redistribution within the volume of the incident laser beam. This interpretation is supported by the comparison of relaxation times. For typical conditions (an absorbed laser power of 1 W/cm^2, a gas pressure of $p(SiH_4) = 10$ mbar and a gas temperature of 10^3 K) the excitation rate for a molecule is about $W_e \approx 10^3$ s^{-1}; and the average time between collisions about 10^{-8} s. The vibrational-translational relaxation rate is about $1/\tau_{v-T} \approx 10^4$ s^{-1}. Therefore, (2.33) seems to be readily fulfilled. Nevertheless, the detailed dissociation mechanism is still under discussion ⌊2.116-119⌋.

In spite of the fact that in LCP most reactions induced by infrared laser light are nonselective, there are some clear exceptions. Among these are gas-phase etching reactions (Chap.8) that are based on multiphoton vibrational excitation and dissociation (MPD) of precursor molecules such as SF_6, CF_3Br, CDF_3. It is SF_6 that has been most extensively studied with respect to both its fundamental excitation mechanisms ⌊2.58-60,109,110,112-114, 121-128⌋ and its etching characteristics ⌊2.129-133⌋ under pulsed CO_2 laser irradiation. For low laser fluences ranging from about 0.1 to 1 J/cm^2 (2-20 MW/cm^2) non-dissociative coherent excitation occurs according to

$$SF_6 + nh\nu(CO_2) \longrightarrow SF_6^* , \qquad (2.34)$$

where n may be greater than 3. The * indicates the vibrational excitation of the molecule. The difference between SF_6 and SiH_4 in the initial stages of absorption is probably due to the extremely dense rotational structure that generally makes compensation of anharmonicity effects in heavy polyatomic molecules more probable. In other words, contrary to SiH_4, pumping of SF_6 into the quasi-continuum is possible without intermediate collisions. On focusing the laser light to power densities of 5 to 10 J/cm^2, multiphoton dissociation of the SF_6 molecule is observed. This may be symbolically described by

$$SF_6 + Nh\nu(CO_2) \longrightarrow SF_5 + F , \qquad (2.35)$$

where N is usually 30 or greater. SF_5 is unstable and further decomposes into SF_4 and another F atom.

Multiphoton absorption spectra show a number of characteristic features: a distinct resonance behavior, a broadening and shifting of the resonance to lower frequencies with increasing laser fluence and a strong dependence of the intensity on laser fluence. For SF_6, these characteristics have been studied by BAGRATASHVILI et al. ⌊2.125⌋. They compare favorably with laser etching experiments performed in a SF_6 atmosphere (Chap.8). A further point to consider for selective infrared LCP is the dependence of the dissociation yield on gas pressure (see e.g. ⌊2.52⌋ and references therein). For many monomolecular gases, the dissociation yield is independent of the gas pressure, within a certain range. In the case of SF_6 this has been demonstrated for the range 0.1 mbar $\leqslant p(SF_6) \leqslant$ 5 mbar [2.135]. This behavior is related to the fact that the v-v exchange between molecules of the same type can take place without a reduction of the average vibrational energy. Collisions between *different* types of molecules can result in a decrease or an increase in dissociation yield. For SF_6 [2.124], CF_3I [2.134], etc., an admixture of monoatomic buffer gases decreases the dissociation yield. For other molecules such as CDF_3, C_2H_4, $C_2H_2F_2$, however, the dissociation yield shows a pronounced maximum when the buffer gas pressure is increased. In the case of CDF_3 [2.135] with Ar, this maximum occurs at a pressure of $p(Ar) \approx$ 25 mbar and exceeds the monomolecular yield by a factor of about 45. The influence of admixtures on the dissociation yield depends on the types of interacting molecules, the gas pressure, the laser fluence, etc. Vibrational energy transfer requires energy matching between vibrational modes, and will be more efficient with complex polyatomic molecules than with simple molecular or monoatomic species.

3. Laser-Induced Chemical Reactions

Laser light can induce chemical reactions either homogeneously within the gas or liquid phase, or heterogeneously at molecule-solid or solid-solid interfaces. In LCP, homogeneously activated reactions are induced near substrate surfaces with the laser beam at parallel incidence (see Fig.4.3). The photoproducts that diffuse to the substrate surface may give rise to extended thin film deposition (Sect.5.3), or to *nonlocal* etching of the material surface (Chap.8). Heterogeneously activated reactions are generally performed at normal incidence of the laser light. As shown schematically in Fig.3.1, such reactions can take place in adsorbed layers, at gas-solid or

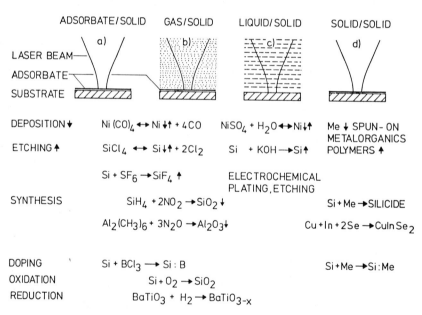

Fig.3.1. Examples of laser-induced chemical reactions at interfaces. For simplicity, not all reaction products are included in the formulas. The arrows refer to deposition ($\downarrow$) and etching ($\uparrow$). Me stands for metal. $\longleftrightarrow$ means that the reaction can be turned around by simply shifting the chemical equilibrium to the other side

liquid-solid interfaces or within the surface of the material itself. Examples of such reactions that result in local material deposition, etching, surface modification or compound formation are included in the figure. Clearly, homogeneous and heterogeneous laser-induced chemical reactions may be activated both pyrolytically and/or photolytically, depending on the interaction mechanisms discussed in Chap.2. Nevertheless, the reaction rates and pathways can be quite different for homogeneous and heterogeneous reactions. In part this originates from the influence of additional light-molecule-surface interactions: Surface adsorption may change the cross section for photodissociation of molecules by several orders of magnitude. Solid surfaces may promote decomposition of partly (homogeneously) dissociated molecules. Laser light may change the properties of surfaces by lattice phonon excitations, electron-hole pair generation, electron emission, etc.

In this chapter we shall outline some fundamentals of the reaction kinetics, mainly for heterogeneously activated reactions, and discuss some of the additional interaction mechanisms mentioned above.

In general, laser-induced reactions consist of a number of consecutive steps:

1. Transport of reactants into the reaction volume
2. Adsorption of one or more reactants onto the substrate
3. Pyrolytic or photolytic activation of molecules near or at the substrate surface
4. Transport of product atoms or molecules to the surface, with possible recombination or secondary reactions on the way
5. Condensation or further reactions of the products of 3 on the surface
6. Desorption of reaction products from the substrate
7. Transport of reaction products out of the reaction volume

Clearly, in different types of LCP, one or more of these steps will either not occur at all or else will differ significantly. The rate-limiting step will depend on the activation mechanisms (photothermal and/or photochemical), the type of reaction (Figs.3.1a-d), the reaction volume, the density of the reactant medium, the physical and chemical properties of the substrate material, the laser power, wavelength, irradiation time, etc.

The different types of reactions shown in Fig.3.1 will now be discussed in Sects.3.1-4 in more detail. Section 3.5 deals with the spatial confinement of laser-induced heterogeneous reactions.

3.1 Adsorbed Layers

Laser-induced chemical processing may be strongly influenced by the adsorption of reactants, reaction products or impurities. Adsorbates may control reaction rates, the spatial resolution of structures and, in the case of deposition or compound formation, nucleation times and the composition and morphology of films.

Molecules on solid surfaces may be adsorbed either physically or chemically. The bond energies of physisorbed molecules are typically 1-10 kcal/mole (binding energies ≈ 0.05-0.5 eV). For chemisorbed molecules the bond energies are of the same magnitude as those of intramolecular bonds, and typically range from 10 to 100 kcal/mole (≈ 0.5-5 eV). The statics and dynamics of the interactions between atoms or molecules and solid surfaces have been extensively studied ⌊3.1-5,38⌋. Most of these investigations were performed under ultrahigh-vacuum (UHV) conditions. Here, atom- or molecule-surface interactions are studied for low surface coverages or for particle beams at physically well-defined solid surfaces. In this section we will briefly mention only those aspects that are relevant to LCP: The number of molecules being adsorbed depends on the energy with which they are bound to the substrate surface, on their interaction with each other (e.g. by dipole-dipole coupling), on the substrate temperature and on the molecular density of the surrounding medium. The strength of these interactions also determines the extent of changes in the electronic and vibrational properties of adsorbed molecules with respect to free molecules. In any case, adsorption may result in shifts and broadenings of electronic and vibrational energy levels and in relaxation of selection rules for interaction with light. Therefore, the cross section for photoexcitation at a particular wavelength may differ significantly for adsorbed and gas-phase molecules ⌊3.4-6⌋. Contrary to investigations on basic molecule-surface interactions, LCP is generally performed only in high-vacuum (HV) reaction chambers that can be pumped out to between 10^{-6} and a few 10^{-7} mbar. Additionally, substrates are in most cases only chemically cleaned according to standard procedures. As a consequence, substrate surfaces may be contaminated with water vapor, organic molecules, etc., which may change the aforementioned interactions and photochemical properties of adsorbed reactant molecules.

In the following we first discuss adsorbate-adsorbent systems in a vacuum (Fig.3.1a) and then those in a dynamic equilibrium with a surrounding gaseous atmosphere consisting of the gaseous form of the adspecies (Fig.3.1b).

In the first case, the substrate is first exposed to the gaseous reactant, which, after some time of influence, is pumped off. Laser light irradiation

38

may result in selective electronic or vibrational (intramolecular bond or adsorptive bond) excitations of the molecules and/or in charge transfer reactions within the adsorbate-adsorbent complex. For a particular molecular species the excitation probability is determined not only by the physical properties of the substrate, but also by its microscopic surface morphology, e.g. its roughness. For example, a strong increase in surface excitation may be observed when a molecular resonance overlaps with optically active surface resonances such as surface plasmons ⌊3.7,8⌋. The influence of the surface morphology on light-molecule-surface interactions has been clearly demonstrated, e.g. in surface enhanced Raman scattering (SERS) ⌊3.4⌋. Besides direct excitation, laser light may also interact with adsorbates indirectly via local substrate heating. Due to direct or indirect interaction with laser light, adsorbed molecules may desorb from the surface, migrate across the surface, change the nature of bonding to the surface (e.g. from physical to chemical), diffuse into the bulk, or decompose at or react with the solid surface ⌊3.4,5,7-15⌋. The latter cases result in deposition, etching, doping, or compound formation. Examples of some reactions are included in Fig.3.1a.

For pyrolytic laser-induced processes the reaction rate achieved within the adlayer is given by

$$W(r,t) = A_{ad} \, N_{ad}(T(r,t)) \, \exp \lfloor -\Delta E/RT(r,t) \rfloor \, , \qquad (3.1)$$

where A_{ad} is a constant, N_{ad} the temperature-dependent number of adsorbed molecules per unit area and ΔE the apparent chemical activation energy that characterizes the slowest step in the chain of chemical reactions involved. The temperature distribution is given by $T(r,t) = T_0 + \Delta T(r,t)$. Here, T_0 is the overall substrate temperature and ΔT the laser-induced temperature rise (Sect.2.1).

For single-photon photolytic processes, the reaction rate achieved within the adlayer can be described by

$$W(r) = B_{ad} \, \frac{N_{ad}(T_0) \sigma_{ad} \, P}{h \nu w_0^2} \, \exp(- \, 2r^2/w_0^2) \, , \qquad (3.2)$$

where B_{ad} is again a constant, σ_{ad} the dissociation cross section of the adspecies, w_0 the radius of the laser focus and P the effective laser power (Sect.4.2). The value of N_{ad} may be decreased with respect to the unirradiated case due to bond-selective excitation that may result in photodesorption ⌊3.12-15⌋.

It is clear that because of the small number of molecules adsorbed on the solid surface, the amount of deposited, etched or doped material is very small. Because of the exponential decrease of N_{ad} with temperature, this is even more pronounced when the incident laser light heats the substrate. On the other hand, molecules may diffuse along the surface and may thereby modify the effective factor N_{ad} in (3.1) and (3.2). The diffusion length within the time τ is given by $l \approx (2D\tau)^{1/2}$ (the exact result depends on the geometry and size of the reaction zone) with the diffusion coefficient $D = D_0 \exp(-\Delta E_d/RT)$. In the initial phase of laser irradiation, replenishment of reactant molecules may occur quickly by surface diffusion from nearby regions; therefore depletion of reactants will be unimportant, and the reaction rate is mainly determined by the photodecomposition rate within the reaction zone. At later times, however, reactant molecules must diffuse from further away, and ultimately the rate at which this diffusion occurs determines the reaction rate. Clearly, these regimes will also depend on the incident laser power. Examples of LCP from adlayers are given in Chaps.5 and 8.

The role of adsorbed layers may be quite different in laser-induced chemical processing from the gas phase. Because such experiments are usually performed at pressures ranging from 10^{-2} mbar up to more than 10^3 mbar, multiple-layer molecular films are formed on top of the more strongly bound first molecular monolayer. The total number of adsorbed molecules depends on the strength of bonding, the temperature, and the gas pressure. In gas-solid systems where the gas pressure p is near to the vapor pressure p_0, the density of such layers is similar to that of the condensed phase. An empirical equation that provides a good fit to the isotherms of many physisorbed systems having a uniform temperature T ⌊3.1,2⌋ is

$$\Theta(p,T) = cp/\{(p_0-p)\lfloor 1 + (c-1)p/p_0\rfloor\} , \tag{3.3}$$

where $\Theta(p,T)$ is the amount of adsorbed material with respect to a monolayer coverage. In other words, the total number of adsorbed particles is

$$N_a(p,T) = \Theta(p,T)N_L , \tag{3.4}$$

where N_L is the number of adsorption sites on the substrate. In (3.3), c is a temperature-dependent constant. From (3.3) it is evident that the surface density varies (unlike the gas-phase density) nonlinearly with pressure. The surface coverage is also a sensitive function of temperature, primarily through the variation of the vapor pressure p_0 with temperature. The change in Θ with even slight changes in temperature is a consequence of the relatively weak binding of physisorbed layers.

Laser radiation may again interact with the adsorbed species as already described above. The coverage of the reactant - and thereby the reaction rate - may be changed if the laser light selectively desorbs one of the components involved in the chemical reaction. Additionally, electronic or vibrational excitation of gas-phase species may change their sticking coefficient. For example, halogen radicals, produced by photodissociation of the corresponding gas-phase molecules, strongly physisorb or chemisorb on semiconductor surfaces, while the corresponding parent molecules are only weakly bound. Another example is the increase in sticking coefficient of vibrationally excited SF_6^* molecules with respect to SF_6 in the vibrational ground state, on semiconductor surfaces (see Sect.2.2 and Chap.8). Adsorption may also change significantly as a result of nonthermal photoexcitation of the substrate surface. Different mechanisms of this type are outlined in Sect.3.2 and Chaps.5-8.

Because of the great thickness of adlayers which are in a dynamical equilibrium with the surrounding gas phase, their influence on reaction rates can become important and, under certain circumstances, dominant.

3.2 Gas-Solid Interfaces

Irrespective of whether a chemical reaction is activated by mainly pyrolytic or mainly photolytic mechanisms, the reaction rate at low laser powers is determined by the kinetics, while at higher laser powers mass transport is rate limiting.

3.2.1 Kinetically Controlled Region

In pyrolytic LCP, the reaction rate in the kinetically controlled region is

$$W(p_i, T(r,t)) = A_g k_0(p_i, T) \exp\lfloor -\Delta E/RT(r,t) \rfloor , \qquad (3.5)$$

where p_i are the partial pressures of the reactants and reaction products, and ΔE is the apparent chemical activation energy. The pre-exponential factor k_0 also depends, although less strongly, on temperature and, additionally, on the partial pressures p_i ⌊see e.g. (5.2)⌋; A_g is a constant. Equation (3.5) also applies to homogeneously induced gas-phase reactions. Then T is the laser-induced gas-phase temperature. In the following, however, we shall consider only heterogeneous reactions at gas-solid interfaces. In this case,

T(r,t) is the laser-induced temperature distribution within the processed area of the solid (Sect.2.1). For flat structures, the thickness of the deposited material, or the depth of the etched or transformed pattern, is obtained by integrating (3.5)

$$h(r,t_i) = A_g' \int_{t_n}^{t_i} dt \; k_o(p_i,T(r,t)) \; \exp\lfloor -\Delta E_j/RT(r,t)\rfloor \; , \tag{3.6}$$

where t_i is the laser beam illumination time, and t_n the latent time before the reaction commences. It is important to note that the temperature distribution $T(r,t)$ is an implicit function of the geometry $h(r,t)$, which influences the heat transport from the processed region. The change in temperature during the time t_i may result in a change in the pathway of the chemical reaction, which is then characterized by a different apparent chemical activation energy. This is indicated by the index j in ΔE. In direct writing of patterns, the integral in (3.6) must be extended from $-\infty$ to $+\infty$. However, in most cases, $t_i - t_n$ can be approximated by the dwell time of the laser beam. The absolute value of r in (3.6) is the distance from the center of the cross section of the pattern, while the direction of r is parallel to $v_s \times k_L$ (v_s and k_L are the scanning velocity and the wave vector of the laser beam; as before, we indicate vectors by arrows only if necessary). The influence of scanning on the temperature distribution is not negligible in LCVD, even for small velocities, if $\kappa_D/\kappa_S > 1$. This is due to the interdependence between the geometry of the deposit (which in turn depends on v_s, as shown for example in Fig.5.14) and the temperature distribution, which becomes all the more pronounced with increasing ratio κ_D/κ_S (see Sect.2.1.2). The width of structures can be calculated in analogy to (3.6).

The confinement of the laser-induced temperature distribution, and thereby of the chemical reaction, in pyrolytic microchemical processing, causes some significant differences in comparison to standard large-area processing techniques ⌊3.16⌋. First, pyrolytic LCP can be performed at much higher partial pressures of reactants. This results in reaction rates that are several orders of magnitude higher than in standard large-area techniques. For example, pyrolytic LCVD can be performed at pressures of up to at least 10^3 mbar (Sect.5.2), while in standard CVD practical partial pressures reach only a few millibars. Due to the uniform heating of the substrate in CVD, higher partial pressures would cause gas-phase nucleation. This would result in uncontrolled deposition. Second, in microchemical processing, three-dimensional diffusion of molecules to and from the reaction zone becomes effective, while in large-area processing only the component normal to the

42

substrate is relevant. Therefore, transport limitations in LCP will arise only at much higher temperatures, i.e. (3.5) will be valid over a wider temperature range. Clearly, these differences from large-area techniques become more significant as the temperature distribution becomes more localized. Finally, it should be mentioned that the aforementioned advantages of microchemical processing also apply, with some restrictions, to laser-induced large-area processing with a line focus (Fig.4.2b).

We now turn to photolytic gas-phase processing. For negligible divergence of the laser beam away from the substrate, we can assume cylindrical symmetry (this holds as long as the Rayleigh length is much greater than w_0; see Chap.4.2). Then, the density of excited atoms or molecules is given by

$$\rho = (N_g \sigma_g / h\nu)\, I(r') \, , \tag{3.7}$$

where σ_g is the gas-phase excitation or dissociation cross section, N_g the number of reactant gas molecules per volume and $I(r')$ the laser intensity at r' within the gas phase. The reaction rate at a point r on the substrate surface due to gas-phase decomposition is obtained by integrating the contributions from the decomposition in the entire gas volume above the substrate $\lfloor 3.17,18 \rfloor$

$$W(r,w_0) = B_g' \, \frac{sN_g\sigma_g P}{h\,\nu w_0{}^2} \int d\phi_a \int dz \int r'dr' \, \lfloor z/d^3(r) \rfloor \, \exp\lfloor -2r'^2/w_0{}^2 \rfloor \, . \tag{3.8}$$

Here, s is the sticking coefficient of the species incident on the substrate, $d(r)$ is the distance of the volume element dv within the gas phase from the point r on the substrate and ϕ_a is the corresponding azimuth angle. Note that (3.8) does not take into account desorbed species that are not deactivated $\lfloor 3.19 \rfloor$. In the center of the laser focus, i.e. for $r = 0$, (3.8) becomes

$$W(0,w_0) = B_g \, \frac{sN_g\sigma_g P}{h\,\nu\, w_0} \, . \tag{3.9}$$

Comparison of (3.9) and (3.2) shows that the photolytic reaction rates for adsorbed- and gas-phase molecules are $W \propto w_0^{-2}$ and $W \propto w_0^{-1}$, respectively. The additional factor w_0 in (3.9) is due to the fact that species created at distances larger than w_0 are distributed over such a large area that they do not contribute appreciably to the deposition rate in the center of the laser focus. These equations also show that in photolysis gas-phase processes should depend linearly on pressure (if we neglect deactivation of species by collisions), while surface-phase processes should follow adsorption isotherms, which depend nonlinearly on pressure. Surface-phase processes are sensitive to temperature variations, especially near the gas-liquid coexistence curve, due to variations in the surface coverage Θ. Gas-phase processes, on the other hand, are not very sensitive to temperature variations, at least not in the linear growth regime. A further point is that photochemical reactions that involve subsequent collisional reaction partners will generally have reaction orders greater than unity. This means that the molecular densities and coverages appearing in (3.9) and (3.2) must be raised to effective powers greater than unity.

Moreover, it is often difficult to reveal the temperature dependence of the reaction rate and thereby the dominant activation mechanism. In such cases, it is often easier and more enlightening to investigate the wavelength dependence of the reaction rate. For materials with constant absorbance in the spectral region under investigation, such as, for example, metals within the visible spectral range, the total laser power absorbed by the substrate, and therefore the temperature distribution, is independent of the laser wavelength. In this case, the rate of a *thermally* controlled reaction at a gas-solid or gas-liquid interface should remain unchanged.

3.2.2 Mass Transport Limited Region

At high laser intensities, the fundamental limits on heterogeneous reaction rates are determined by mass transport. For the pressures normally used in gas-phase LCP, the ultimate rate is determined by a balance between the gas-phase transport by diffusion of reactants and products into and out of the reaction volume. Here, we disregard surface diffusion of species and convective flow. For large-area planar reactions, this balance is obtained by solving the one-dimensional diffusion equation. As outlined above, the situation is different in microchemical processing. Here, the three-dimensional character of the diffusion of reactants and reaction products becomes relevant. In the following we assume a hemispherical geometry for the zone of diffusive molecule transport, as shown schematically in Fig.3.2. The reaction rate is then proportional to the concentration of reactants within this zone. The concentration outside this zone can be assumed to have a constant value, n_∞.

In the simple case where reactant diffusion is slower than product diffusion, such as in the case of SiH_4 decomposition into Si and H_2, the time-dependent surface-reaction flux, $j(t)$, is given by [3.20]

$$j(t) = \frac{2Dn_\infty}{r_0 + d} \{1 + \frac{d}{r_0} \exp\lfloor (\frac{1}{r_0} + \frac{1}{d})^2 4Dt \rfloor \, erfc\lfloor (\frac{1}{r_0} + \frac{1}{d})(4Dt)^{1/2}\rfloor\} \, , \quad (3.10)$$

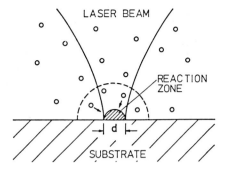

LASER BEAM

REACTION ZONE

SUBSTRATE

Fig.3.2. Zone of diffusive transport of molecules in laser-induced microchemical processing

where D is the molecular diffusivity, which includes the effect of possible carrier gases. Here, for simplicity, D has been assumed to be constant. In (3.10), $r_0 \equiv 4D/\eta'v_m$ is a length scale proportional to the mean free path in the gas, η' is the reaction efficiency per surface collision, and v_m is the rms velocity of reactant molecules near the surface. Figure 3.3 shows the time evolution of the reaction flux for various values of the reaction zone radius. The conditions chosen are representative for laser-induced CVD and gas-phase etching. The time required to approach steady-state conditions can be specified by

$$\tau_0 \approx \frac{r_0^2 \, d^2}{4D(d + r_0)^2} \, . \tag{3.11}$$

Because $D \sim p^{-1}$, the time τ_0 increases linearly with pressure for small spot sizes, i.e. if $d \ll r_0$, and becomes inversely related to pressure for large spots, i.e. if $d \gg r_0$. The equilibration time τ can be < 10 ns for small spot sizes and moderate pressures (< 1000 mbar). This means that for small spots even pulsed LCP is often in a steady-state condition. Therefore, in

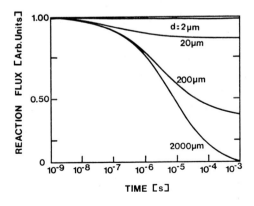

Fig.3.3. Time dependence of the molecular flux into a microreaction zone of width d. p = 133 mbar, T = 1273 K, D = 2.8 cm^2/s, η' = 0.1 (after [3.20])

45

most cases, deposition or etching is determined by the steady-state solution of the diffusion equation.

The steady-state surface reaction rate is given by

$$j(t \longrightarrow \infty) = 2Dn_\infty /(r_0 + d) \ . \qquad\qquad (3.12)$$

If $d \ll r_0$, this rate approaches $2Dn_\infty /r_0 = \eta'v_m n_\infty /2$, i.e. it becomes nearly proportional to pressure, but independent of d. If $d \gg r_0$ the rate saturates and approaches $2Dn_\infty /d$, i.e. it becomes independent of pressure, but decreases with d. The pressure dependence of the steady-state reaction rate is plotted in Fig.3.4 for various radii of the reaction zone. The values chosen are typical for pyrolytic gas-phase processing. The fluxes predicted, e.g. for $d \approx 2$ μm and a pressure of 500 mbar, are about 10^{22} molecules/cm^2 s.

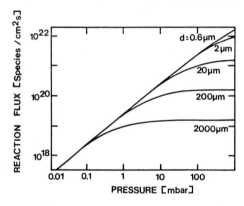

Fig.3.4. Steady-state values for the molecular reaction flux as a function of gas pressure at times t $\gg \tau_0$. $D = \alpha'/p$ with $\alpha' = 2.7 \cdot 10^2$ mbar cm^2/s, $r_0 = \beta'/p$ with $\beta' = 6.7 \cdot 10^{-2}$ mbar cm, T = 1273 K, $\eta' = 0.1$ (after [3.20])

This corresponds to a deposition or etch rate of several 10^3 μm/s. This rate is sufficient to explain the tremendous deposition and etch rates achieved in laser-induced microchemical processing (Chaps.5 and 8). It should be mentioned that for a comparison of reaction rates that can be achieved in laser-induced microchemical processing and in standard large-area techniques, one must take into account the fact that standard processing is often performed in flushed reactors. In such cases, diffusive transport need only occur across a thin boundary layer that has a typical width of $\delta \approx 3$ mm. The enhancement in laser microchemical processing is of the order of $2\delta/d$; for $d = 2$ μm, this factor is 10^3. It should be noted that at very low partial pressures of reactant molecules the dominant mass transport mechanism need not be gas-phase diffusion, but can instead be diffusion of adsorbed molecules along the substrate surface (Sect.3.1). At very high molecular densities, on the other hand, mass transport by convection may become important (Sect.3.3).

The reaction rates achieved in gas-phase LCP are often well below the transport-limited rates discussed above. This is especially true in photolytic processing when the laser light is absorbed within the gas phase. Here, the rate is commonly restricted by gas-phase nucleation or recombination (see Chaps.5 and 8). Gas-phase nucleation or recombination, however, can only be controlled by reducing the atomic or molecular density. Sometimes the same effect can be achieved by inert-gas buffering. In any case, this results in a lowering of deposition or etch rates. In photolytic LCP the reaction rates may be even further limited by the available laser power and/or the mismatch of the laser wavelength and the maxima in the absorption cross section of reactant molecules.

3.2.3 Solid Surface Excitations

So far we have discussed reaction rates based on laser-induced substrate heating or on direct photochemical dissociation of adsorbed- or gas-phase species. There are, however, a great number of other effects that may significantly influence laser-induced reaction rates at gas-solid interfaces. Among those are autocatalytic effects and dissociative photoelectron capture ⌊3.21⌋. Both effects have been studied in connection with metal deposition ⌊3.22,23⌋. In semiconductors and insulators, the excitation of the band gap may cause a nonthermal increase in reaction rates. This effect has been studied in some detail in connection with dry- and wet-etching of Si and GaAs (Chap.8). Another effect, which has already been mentioned in Sect.3.1, is the electromagnetic field enhancement on metal and semiconductor surfaces ⌊3.7-10,24,25⌋. This effect can significantly alter reaction rates in gas-phase deposition, surface modification and etching reactions. Strong surface electric fields induced by the incident laser radiation may significantly increase surface adsorption and diffusion of species into the solid surface ⌊3.24-26⌋. This mechanism has been proved to be particularly important in connection with surface oxidation and etching reactions in a halogen atmosphere (Sects.6.1 and 8.2). While some initial studies of these different mechanisms have been performed already, detailed investigations are still lacking. Furthermore, it should be emphasized that a separate discussion of single effects greatly oversimplifies the situation of chemical reactions at gas-solid interfaces. A more satisfactory understanding requires consideration of chemical phenomena in the gas, adsorbed and solid phases and, in addition, the often subtle interactions between these three phases. Furthermore, as already pointed out in various sections, during deposition, doping, etching, etc. the morphology and physical properties of the surface

47

will change. Such changes in surface properties, will, in turn, influence all chemical and other mentioned effects near the interface.

3.3 Liquid-Solid Interfaces

Many of the mechanisms and results outlined in the foregoing sections are also relevant to liquid-phase processing. The microscopic mechanisms of photoreactions can again be based on mainly pyrolytic or mainly photolytic effects. Deposition or etching may be controlled by the chemical kinetics or by mass transport. In the kinetically controlled region the mechanisms and the corresponding equations are analogous to those outlined above. Similarly, mass transport by diffusion can be described by (3.10) and (3.12) for the time-dependent and the steady-state cases, respectively, if the effects of the liquid solvent are included in the molecular diffusivity D. However, in dense media additional mass transport mechanisms often become important. In pyrolytic LCP, for instance, the strong temperature gradients can induce convection, turbulence and bubbling.

   Another type of LCP, which involves mechanisms quite different from those discussed up to now, is laser-enhanced electrochemical plating and etching, where the substrate is externally biased with respect to a counterelectrode (Sect.4.1). Detailed experiments by VON GUTFELD ⌊3.27,28⌋ have shown that the enhancement of the reaction rate is based on local laser-induced heating. Photochemical effects within the liquid have been excluded (see Sect.5.2.3 and Chap.8). We shall therefore concentrate on the question: what is the effect of local heating on the charge transfer and mass transfer rates within an electrochemical system? To answer this question, we have to consider the temperature dependence of the current density. In the kinetically controlled regime, i.e. at low overpotentials, the current density is given by the Butler-Volmer equation ⌊3.29⌋

$$i = i_0 \{\exp\lfloor(1-\beta)\xi zF'/RT\rfloor - \exp(\beta\xi zF'/RT)\} \, , \tag{3.13}$$

where β is an exchange coefficent, ξ the overpotential, z the number of charges per ion undergoing reduction and F' the Faraday constant. The exchange current density i_0 (also called the charge transfer rate) is given by

$$i_0 = (znF'k_BT/h) \exp\lfloor-(\Delta E + \beta zF'\Delta\Phi_e)/RT\rfloor \, , \tag{3.14}$$

where n is the ion concentration per square centimeter at the electrode and

48

$\Delta\Phi_e$ the equilibrium potential. The temperature dependence of i_0 is dominated by the Arrhenius term $\exp(-\Delta E/RT)$, since $\Delta E > \beta zF'|\Delta\Phi_e|$. Therefore, the rate always increases with increasing temperature. The equilibrium potential $\Delta\Phi_e$ is also temperature dependent. For many electrolytes, $\Delta\Phi_e$ shifts to larger (more positive) values with increasing temperature. This has been observed, for example, with Au and Cu plating. However, there are systems where the opposite temperature dependence of $\Delta\Phi_e$ has been found [3.27,28].

At higher overpotentials, mass transport of ions becomes rate limiting. In addition to diffusion and forced convection, the mass transport due to the gradient of the electric potential Φ must be taken into consideration. According to Nernst and Planck, the mass transport equation can be written in the form

$$j_i = D_i \nabla n_i + (z_i F'/RT)\, D_i n_i \nabla\Phi - n_i v \ . \tag{3.15}$$

Forced convection [last term in (3.15)] has been estimated for liquid-phase LCP by solving the three-dimensional Navier-Stokes equations together with the appropriate boundary conditions [3.30]. At higher laser light irradiances, an additional rate enhancement due to localized boiling was observed [3.27,28]. Another possible way of increasing the mass transport of ions was demonstrated in laser-enhanced jet-plating (see [3.31] and Sects.4.1 and 5.2.3).

Two types of laser-enhanced plating reactions, which use no external driving force (EMF), have been investigated [3.27,28,32-34]. The first of these is laser-enhanced electroless plating, where the charge balance is maintained via a catalytic reducing agent incorporated in the solution [3.34]. Fast plating was observed within the area illuminated by the laser light. Virtually no plating took place outside that region. Thermobattery or laser-enhanced exchange plating (etching) utilizes the positive (negative) shift in the rest potential, which results in a local thermobattery [3.27,28,32-34]. This results in increased plating (etching) in the heated region and simultaneous etching (plating) in the cooler peripheral regions. The simultaneous plating and etching maintains overall charge neutrality.

Nonthermal mechanisms in liquid-phase processing (without an external EMF) have been revealed in connection with wet-etching of semiconductors at low laser light intensities (see Sect.8.2). Here, the semiconductor is immersed in an ionic solution containing an oxidizing agent. In equilibrium without light irradiation, the Fermi level in the semiconductor must match the redox level of the liquid. As a consequence, the semiconductor bands will be bent upwards or downwards, depending on whether the semiconductor is n- or p-type

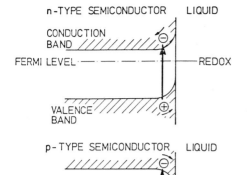

n-TYPE SEMICONDUCTOR   LIQUID

CONDUCTION
BAND
FERMI LEVEL ─ ─ ─ ─ ─ ─ ─── REDOX

VALENCE
BAND

p-TYPE SEMICONDUCTOR   LIQUID

FERMI LEVEL ─ ─ ─ ─ ─ ─ ─── REDOX

Fig.3.5. Band bending of n- and p-type semiconductors at the interface with a liquid electrolyte. Conditions are similar at gas-semiconductor interfaces. Charge carrier flow under illumination with band-gap radiation is indicated

(see also ⌊3.35⌋). This is shown schematically in Fig.3.5. The band bending can be easily understood: The oxidizing agent will attract electrons. For n-type material, a negatively charged surface layer is thereby formed. This causes a strong upward band bending resulting in a barrier preventing any further flow of electrons towards the suface. In p-type material, on the other hand, electrons are the minority carriers. The Fermi level will be too low to permit substantial electron transfer. In this case, bands are bent downwards, creating a barrier against any further migration of holes. Light illumination, at an energy that exceeds the band gap, will generate electron-hole pairs. In n-type material the holes will drift to the surface, the electrons further into the bulk. The holes can be considered as ionized or broken bonds within the surface. The disrupted lattice will strongly interact with the negatively charged surface species. This may result in oxide formation. If the solution contains an acid that dissolves this oxide, the semiconductor will dissolve into positive ions. For p-type material there is a depletion of holes near the surface. Therefore, light-induced etching as described for n-type material will occur at a slower rate or not at all. For particular liquid-solid interfaces, the process may even be reversed, resulting in material deposition. In this latter case, however, the change in the physical properties at the interface will, in general, rapidly terminate the reaction.

A theoretical model that predicts the spatial resolution for grating formation on n-type semiconductors by light-enhanced electrochemical etching has been developed by OSTERMAYER et al. ⌊3.36⌋ (see also Sect.8.2.2).

3.4 Solid-Solid Interfaces

In this section we consider laser-induced reactions within solid surfaces or at solid-solid interfaces (case d in Fig.3.1). Such reactions may again be activated photothermally and/or photochemically. They may result in formation or breaking of chemical bonds. Examples are laser-induced surface doping of materials, e.g. semiconductors, from thin evaporated layers (Sect.6.3), material synthesis, e.g. from multilayered structures consisting of the elements (Sect.7.2), or ablative photodecomposition of organic polymers (Sect.8.4). Material doping and synthesis often require surface melting in order to allow liquid-phase diffusion of atoms or molecules. For Si, for example, the liquid-phase diffusion coefficient for many dopant atoms (Sb, Bi, Ga, In) is between 10^{-4} and $7 \cdot 10^{-4}$ cm^2/s, while the solid-phase diffusion coefficient near melting is, typically, between 10^{-6} and 10^{-5} cm^2/s $\lfloor 3.39 \rfloor$. The total doping dose is proportional to the diffusion length, which, in the case of diffusion perpendicular to the substrate surface, is given approximately by

$$1 \approx (2{<}D{>}t_i)^{1/2} \; ; \tag{3.16}$$

where $<D>$ is the molecular diffusion constant averaged over the time t_i. For cw laser processing t_i can be approximated by the dwell time of the laser beam, i.e. $t_i = t_d = 2w_0/v_s$. For pulsed laser processing $t_i = N\tau$, where N is the number of laser pulses of duration τ. However, because of the high material densities, thermal and stress gradients can considerably increase material diffusion (see e.g. $\lfloor 3.37 \rfloor$), and thereby the local reaction rates, over the values achieved in large-area isothermal processing. For example, the diffusion flux driven by a temperature gradient normal to the substrate surface is given by

$$j = \frac{n_c \Delta E}{k_B} \left\langle \frac{D}{T^2} \frac{\partial T}{\partial z} \right\rangle \; , \tag{3.17}$$

where n_c is the local dopant concentration and ΔE the activation energy of a lattice vacancy (typically about 2 eV). The brackets $< >$ again denote the time-average. Correspondingly, the high heating and quenching rates achieved with laser light may result in stress gradients that cause a diffusion flux similar to (3.17).

Laser-induced bond breaking between surface atoms may result in surface modification or in ablation of the material. Ablation may be simply based on laser-induced melting and/or thermally activated vaporization of the material. This is generally the case in conventional laser machining (cutting, drilling, etc.) as described in Chap.1. The situation becomes much more complex when a material surface is irradiated with short but very intense laser pulses. Detailed investigations of this kind have been carried out for a great number of materials ⌊3.40-43⌋. In this regime, laser-induced ablation has been termed laser sputtering. Laser sputtering cannot always be explained on the basis of a thermal model only.

Direct nonthermal bond breaking seems to be the dominating mechanism in ablative photodecomposition (APD) of organic polymers and biological materials under ArF laser irradiation ⌊3.44-50⌋. For wavelengths λ < 200 nm, organic polymers have a very high absorption coefficient, which is typically of the order of $5 \cdot 10^3$ to 10^5 cm^{-1}. Therefore, the absorbed energy is essentially deposited within a 2 to 0.1 μm thick layer. The UV radiation may excite the organic molecule to above the dissociation limit (Fig.2.15b) or to an upper bound state from which, due to crossing into another unstable or less stable electronic state, dissociation can occur (Figs.2.15c,d). For wavelengths λ < 200 nm, the dissociation of bonds is extremely efficient, typically larger than 50%. Nevertheless, intersystem crossing or internal conversion can also lead to electronic deexcitation via vibrational relaxation and thereby to a dissipation of the excitation energy into heat. Ablation of polymers with visible and IR radiation has been demonstrated as well. Here, melting and vaporization of the material seems to be the dominating mechanism.

Laser ablation of organic polymers has been modeled for both photochemical and photothermal processes ⌊3.44-50⌋. According to the model of GARRISON and SRINIVASAN ⌊3.48⌋, photochemical ablation will occur via fragmentation of the polymer into monomers, which, because of their larger specific volume, interact among themselves and with the remaining material via a repulsive potential. This results in a volume explosion (on a picosecond time scale) without melting. The reaction products will be ejected from the photolyzed area at a very high speed (estimated as 10^5-$2 \cdot 10^5$ cm/s) and into a small solid angle within about 30° of the surface normal. For photon energies less than typical bond energies (which are about 6.0 to 2.5 eV for the materials under consideration), corresponding to wavelengths λ > 200 to 500 nm, single-photon dissociation becomes impossible. Thus, in this wavelength region, ablation may occur either via sequential multiphoton excitation or via heating, melting and evaporation. In the latter case, the predicted angular distribution of the ablated material becomes very broad, out to more than 60° from the surface normal. While many characteristics of APD of organic polymers (Sect.8.4) support a nonthermal bond-breaking mechanism, there are other observations that suggest a mostly statistical thermodynamic process, with transient melting but without complete energy randomization ⌊3.51,52,55⌋. However, it seems to be clear that the dominating mechanisms depend on the specific material under investigation, as well as on the laser fluence, wavelength and pulse width. In fact, a novel dynamic model has recently been elaborated to allow a consistent analysis of thermal and nonthermal contributions to APD, including the description of fluence thresholds, incubation pulses, fast intra-pulse etching etc. ⌊3.45⌋.

The processes outlined in this section may be performed in a nonreactive atmosphere. Nevertheless, reactive surroundings, for example air or oxygen, may affect the reaction rates and chemistry.

3.5 Spatial Confinement

The spatial confinement of laser-induced chemical reactions at interfaces is determined by

1. The laser beam spot size
2. The spatial confinement of the laser light-induced excitation
3. Laser light-induced disturbances of the surrounding medium
4. The nonlinearity of the surface chemical process itself

The single points will now be discussed in greater detail. Normal incidence of the laser beam onto the substrate is assumed.

For a Gaussian beam, the diffraction-limited diameter of the laser focus is given by $2w_0 = 4f\lambda/\pi a \approx 1.3\ f\lambda/a$ (Sect.4.2). For $f/a \approx 1.5$ optics, we obtain $2w_0^{min} \approx 2\ \lambda$, i.e. for 500 nm radiation, the minimum laser spot size becomes about 1 μm (index of refraction n = 1).

The confinement of the laser-induced excitation within the gas or liquid phase, on the substrate surface and in its bulk, is also of considerable importance. Consider, for example, species that are photoexcited or dissociated within the gas or liquid phase. Such species will randomly diffuse towards the solid surface. Consequently, deposition on, or etching of areas beyond the focal point region will occur. The spatial resolution is thereby decreased. This problem can be overcome, in part, by a careful selection of the processing parameters. In pyrolytic processing from the gas or liquid phase, photothermal or photochemical excitation within the volume of the laser beam can be avoided to a large extent by a proper choice of the laser wavelength and irradiance. In photochemical gas- or liquid-phase

53

processing, confinement of the reaction to the substrate surface would be possible only if the dissociative continuum of the adsorbed molecules shifted considerably towards red, so that proper selection of the laser wavelength would allow adsorbed-phase but not gas- or liquid-phase photolysis (see Sect.3.1). In spite of the fact that a shift of the dissociative continuum of adsorbed species with respect to gas- or liquid-phase molecules plays an important role in many systems investigated in LCP, photoexcitation or dissociation within the total volume of the incident laser beam was observed in essentially all cases investigated in photolytic LCP. However, tight focusing of the laser beam not only limits the area of excitation on the substrate, but also confines the relevant volume of excitation within the adjacent surrounding medium. See Sect.3.2 and (4.3). Species that are generated at a distance larger than their mean free path for deactivation (by collisional recombination with complementary photofragments, or reactions with parent or other molecules) will not reach the substrate. In other words, the flux of species to the substrate is limited to a small region around the laser spot. The rate of recombination depends not only on the type and density of molecules in the ambient medium but also on the incident light flux. In gas-phase processing, the confinement of the reaction volume can sometimes be further enhanced by proper selection of the type of buffer gas and the respective partial pressures of gaseous constituents. Low partial pressures of reactants and the admixture of adequate buffer gases may improve the localization but decrease the efficiency of the photodeposition or etching process.

Apart from homogeneous gas- or liquid-phase excitations, there are additional mechanisms that may decrease the spatial confinement. Among these are: thermal diffusion and diffusion of photoexcited carriers within the substrate or, in the case of deposition, within the already deposited material; diffusion of adsorbed reactant species or of photofragments produced on the substrate surface itself, etc. For the broad range of experimental conditions used (Chaps.5-8) the corresponding diffusion lengths may be smaller, comparable or larger than $2w_0^{min}$. In any case, these different mechanisms may decrease the spatial confinement of processing, depending on the parameters used. Reliable estimates are very difficult for many reasons, including: the difficulty of ensuring a correct treatment of the strong temperature gradients in pyrolytic LCP; the unknown bulk and surface properties of the material within the processed region; the lack of reliable data on the corresponding parameters such as gas, surface, and bulk diffusion coefficients, characteristic lifetimes, etc.

Laser light induced disturbances within the surrounding medium may substantially decrease the spatial confinement of the reaction. Strong temperature inhomogeneities in pyrolytic LCP may change the optical index of refraction, cause convection, turbulence or even bubbling. These effects are especially pronounced in dense media (Sect.3.3). In photolytic LCP, the formation of clusters and the coating of entrance windows may attenuate and scatter the incident light.

From the above considerations, one would expect that the minimum lateral dimensions, d_{min}, of laser light generated structures to be $d_{min} \gtrsim 2 \, w_0^{min}$. On the other hand, it has been demonstrated that laser direct writing allows one to generate structures with $d < 2w_0$ (Sect.5.2). In other words, it is possible to produce patterns with lateral dimensions smaller than the diffraction-limited diameter of the laser focus. This shows the importance of process nonlinearities, which have been discussed in the foregoing sections.

To illustrate the influence of nonlinearities in pyrolytic LCP, we have plotted in Fig.3.6 a temperature profile (full curve) induced by a Gaussian laser beam (dotted curve) on a semiinfinite substrate. The figure also shows the normalized reaction rates, $W(r) \propto \exp\lfloor -\Delta E/kT(r) \rfloor$, for two different chemical activation energies ΔE = 22 kcal/mole (curve a) and ΔE = 46.6 kcal/mole (curve b). These activation energies correspond, for example, to those in laser-induced deposition of Ni and Si from $Ni(CO)_4$ and SiH_4, respectively (Sect.5.2). Note, however, that the influence of the laser-processed feature on the temperature distribution was *not* taken into account here. In laser-induced deposition, this approximation is valid only if the

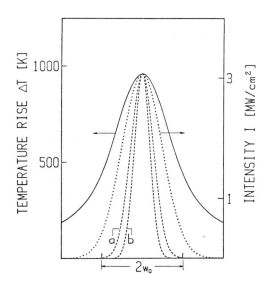

Fig.3.6. Confinement of the chemical reaction in pyrolytic processing. Dotted curve: Profile of the laser beam. Full curve: Temperature rise calculated for κ = 50 W/mK and an absorbed laser power P_{abs} = 0.3 W (surface absorption). The normalized reaction rate is obtained by exponentiating the temperature distribution. Dashed curves refer to apparent chemical activation energies of ΔE = 22 kcal/mole (curve a) and ΔE = 46.6 kcal/mole (curve b)

thermal conductivities for the deposit and the substrate are approximately equal, i.e. if $\kappa_D/\kappa_S \approx 1$ (see Sect.2.1.2). In spite of this restriction in generality, the figure shows some interesting features. The nonlinear dependence of the reaction rate on temperature causes the lateral variation in W to be substantially narrower than in T. The spatial confinement of the reaction increases with increasing apparent chemical activation energy ΔE (compare the dashed curves a and b). When a particular material has to be deposited, however, there is in many cases no choice of parent molecules. One of the exceptions is Si, which can be deposited from either SiH_4 or $SiCl_4$, which have somewhat different activation energies.

The influence of ΔE and of the center temperature will now be studied in more detail. Let us describe the increase in spatial confinement by the ratio w_0/r_e, where $r_e = d_e/2$ is given by the 1/e point in the reaction rate $W(r)$. For temperature-independent parameters, we obtain from (2.7) and (2.10)

$$T(r) = T_0 + \Delta T(r) = T_0 + \Delta T_c \; f(r) \; , \quad \text{with} \tag{3.18}$$

$$f(r) = 2\pi^{-1} \int_0^\infty g(u) \; du \; .$$

The above definition then yields

$$r_e = \tilde{f} \left(\frac{1 - T_0 R(1 + T_0/\Delta T_c)/\Delta E}{1 + T_0 R(1 + \Delta T_c/T_0)/\Delta E} \right) \equiv \tilde{f}(\gamma) \; , \tag{3.19}$$

where $\tilde{f}(\gamma)$ is the inverse function of $f(r)$. In Fig.3.7 we have plotted the ratio w_0/r_e as a function of ΔT_c for various activation energies ΔE. The figure shows that the confinement increases very rapidly with increasing ΔE at low values of ΔE, and much more slowly at higher values. Furthermore, for a certain value of ΔE, there exists a maximal value of w_0/r_e at a certain center temperature $T_c^m = T_0 + \Delta T_c^m$. Clearly, for efficient processing, the practical center temperature, T_c^p, must be well above the threshold temperature for deposition, etching, etc., because otherwise the reaction rate would be extremely small. In many cases, $T_c^p > T_c^m$ and the confinement will be decreased with respect to the optimal value. Besides the mechanisms which decrease the spatial confinement with respect to that expected from Fig.3.7, and which have been discussed above, there are additional fundamental mechanisms that also limit the ultimate resolution of structures. These will be discussed in Sect.5.2.5 in connection with laser direct writing.

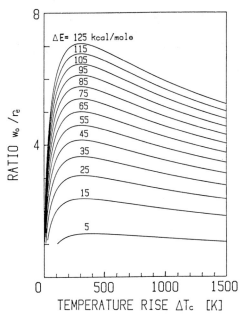

In laser photolysis, based on multiphoton dissociation of adsorbed molecules, the reaction rate is given by $W \propto P^n$, where n is the number of coherently absorbed photons (Sect.2.2). This nonlinearity increases the confinement in photolytic processing.

Whenever nucleation processes are involved in a particular processing step, such as in deposition or compound formation, they will be of great importance for the confinement of the chemical reaction. The reason is the strong nonlinearity of the nucleation process itself ⌊3.54⌋. This nonlinearity is, of course, independent of whether species are generated by pyrolytic or photolytic processes. Nucleation processes can take place within the surrounding medium, as in the gas phase, on the substrate surface or, as in compound formation, within solid films. Nucleation will be discussed in Sect.5.1 in more detail.

4. Experimental Techniques

This chapter is intended to give an overview of the main experimental components and aspects of LCP. Section 4.1 describes different experimental setups for laser microchemical and large-area chemical processing. Section 4.2 comments on lasers that are typically used in LCP. Techniques of measuring deposition and etch rates are outlined in Sect.4.3. Section 4.4 briefly describes different possibilities of in situ temperature measurements.

4.1 Typical Setups

The main components of an experimental setup for LCP consist of a laser and a reaction chamber. The latter contains the reactive species and the substrate. The laser light can be incident perpendicular and/or parallel to the surface of the substrate ⌊4.1,2⌋. The reaction chamber is often operated with a constant flow of the reacting gaseous or liquid species with or without a carrier. In microchemical processing, the reaction chamber can be sealed off in many cases, because of the small amount of species consumed in most of the reactions.

For microchemical processing, the laser beam, in general a cw laser (Sect.4.2), is expanded and then focused at normal incidence onto the substrate surface. A typical setup is shown schematically in Fig.4.1. Laser beam illumination times are electronically controlled with an electrooptical modulator. The eyepiece is used for direct observation of patterns. The position of the objective is optically and electronically controlled (autofocus). Direct writing of micrometer-sized surface patterns is accomplished by translating the substrate perpendicular to the focused laser beam (Fig.4.2a).

Large-area processing can be performed with an experimental arrangement that, apart from some changes in the optics, is similar to that shown in Fig.4.1. The main difference is that the laser beam is scanned over the

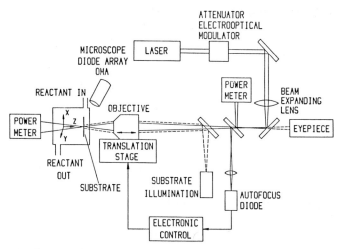

Fig.4.1. Schematic of a typical experimental setup for microchemical processing. The substrate is mounted on an xyz stage. The position of the objective is optically and electrically controlled (autofocus). The eyepiece (or a TV camera together with a monitor) is used for direct observation of patterns

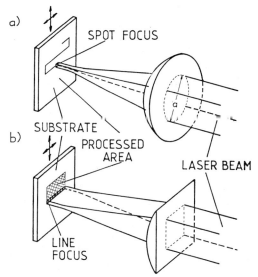

Fig.4.2. Schematic showing the focusing in microchemical (a) and large-area chemical (b) processing in perpendicular configuration (after [4.1])

substrate either unfocused or defocused or focused to a line by means of a cylindrical lens (Fig.4.2b). Another type of experimental setup used in large-area processing is shown in Fig.4.3. Here, a collimated beam is passed, at a certain distance, parallel to the surface of the substrate. Combined parallel and perpendicular irradiation is achieved either by directing the emerging beam onto the surface of the substrate in an arrangement as shown in

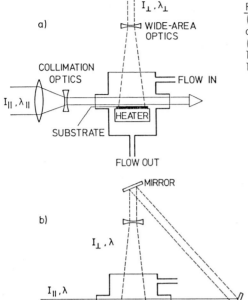

a)

$I_\perp, \lambda_\perp$
WIDE-AREA OPTICS

COLLIMATION OPTICS

$I_\parallel, \lambda_\parallel$

FLOW IN

HEATER

SUBSTRATE

FLOW OUT

b)

MIRROR

$I_\perp, \lambda$

$I_\parallel, \lambda$

MIRROR

SUBSTRATE

Fig.4.3. Optics for parallel (beams shown by solid lines) and combined parallel/perpendicular (beams shown by solid and dashed lines) laser irradiation used in large-area chemical processing

Fig.4.3b, or by using two lasers, perhaps at different wavelengths. In the latter case, one can separately optimize homogeneous and heterogeneous pyrolytic and/or photolytic reactions by proper selections of $I_\parallel$, $\lambda_\parallel$ and $I_\perp$, $\lambda_\perp$, respectively.

In laser-enhanced electrochemical plating (etching), the substrate is negatively (positively) biased with respect to a counterelectrode [4.3,4]. The applied voltages are typically 1-2 V. Figure 4.4 shows an apparatus for laser-enhanced jet-plating, which is the most recent development in this area. In this technique, the mass transport to the substrate is increased by a jet (typical flow velocities are 10^3 cm/s). The laser beam (in most experiments an Ar^+ or Kr^+ laser was used) is focused at the center of the orifice of the jet and is maintained within the liquid column by total internal reflection until impingement on the cathode occurs. In these experiments, the potentiostat is set to deliver constant current, i.e. to plate galvanostatically. In earlier experiments, not using a jet, the substrate was immersed within the liquid electrolyte. Here, the laser beam was directed either through the electrolyte to directly illuminate the electrolyte-substrate interface from the front or, in the case of transparent substrates coated with metal on one side, through the substrate to illuminate

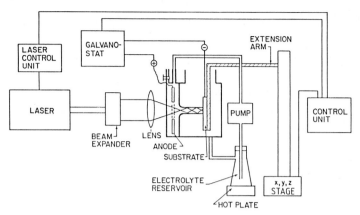

<u>Fig.4.4.</u> Experimental setup for laser-induced electrochemical jet-plating. The laser beam is focused approximately at the center of the jet orifice. The substrate can be moved via the extension arm (after ⌊4.3⌋)

the metal layer from the rear only. These initial experiments simply used a dropping resistor to measure the current flow with and without laser beam illumination. Both continuous and pulsed plating (etching) were demonstrated by modulating the external voltage source, the laser output power, or both synchronously. Both, Ar^+ and Kr^+ lasers were used with power densities ranging from 10^2 to 10^6 W/cm². The plating (etching) mechanisms and experimental results are described in Sects.3.3 and 5.2.3, respectively.

4.2 Lasers

The lasers that are most commonly used in LCP are listed in Table 4.1, together with some of their characteristic features. Micrometer-sized structures are mainly produced with cw lasers, such as Ar^+ or Kr^+ lasers, including frequency-doubled lines. The reason is the tight focusability and good stability of such lasers with respect to the beam profile and the output power. These are prior conditions in direct writing of microstructures with constant and well-defined morphology (Sect.5.2). The TEM_{00} mode, which has been used in most of the experiments, is of Gaussian shape. Then, the incident laser irradiance within the focal plane can be written in the form

$$I(r) = I(0) \exp(-2r^2/w_0^2) \ , \tag{4.1}$$

where w_0 is defined by $I(w_0) = I(0)/e^2$ and is given by approximately (see, for example, ⌊4.5-7⌋)

Table 4.1. Commercial lasers most commonly used in materials processing. Only strongest lines are listed. Data for harmonics are given in parentheses

Laser	Wavelength [nm]	Pulse length	cw Power/ Pulse energy
F_2	157	ns	$\leqslant$2 W; 0.025 J
ArF	193	ns	$\leqslant$10 W; 0.5 J
KrCl	222	ns	$\leqslant$10 W; 0.2 J
KrF	248	ns	$\leqslant$25 W; 1.0 J
XeCl	308	ns	$\leqslant$8 W; 0.5 J
XeF	351	ns	$\leqslant$7 W; 0.4 J
N_2	337	ns	$\leqslant$0.4 W; 0.02 J
HeCd	441.6	cw	0.05 W
Ar^+	351.1 – 363.8	cw	ML 2.5
	454.5	cw	1 W
	457.9	cw	1.2 W
	476.5	cw	3 W
	488.0	cw	7 W
	496.5	cw	2.5 W
	501.7	cw	1.5 W
	514.5 (257.3)	cw	8 (0.1) W
	528.7	cw	1 W
Kr^+	337.4, 350.7, 356.4	cw	ML 1.5 W
	413.1	cw	0.5 W
	476.2	cw	0.5 W
	520.8	cw	1 W
	530.9	cw	1.5 W
	568.2	cw	1.0 W
	647.1	cw	4 W
	676.4	cw	1 W
	752.5	cw	1.5 W
HeNe	632.8	cw	0.05 W
Ruby	694	ns	100 J
Nd:glass	1054	ns, ps	100 J
Nd:YAG	1064	cw,	30 kW
	1064 (532; 354.7; 266)	ns, ps	100 (20; 10; 4) J
CO	5 – 7 μm	cw; μs	0.1 kW; 0.5 J
CO_2	9 – 11 μm	cw; μs, ns	15 kW; 100 J
	R(24) = 10.22 μm		
	P(18) = 10.57 μm		
	P(20) = 10.59 μm		
	P(22) = 10.61 μm		

$$w_0 = 2f\lambda/\pi a \ . \tag{4.2}$$

Here, a is the aperture (Fig.4.2a). It is evident from (4.2) that the diffraction-limited diameter of the laser focus, $2w_0$, decreases with decreasing wavelength. Clearly, w_0 is one of the essential parameters determining the lateral resolution of patterns (Sect.3.5). The length of the laser focus is defined by

$$L = 2z_R = 2\pi w_0^2/\lambda \ , \tag{4.3}$$

where z_R is the Rayleigh length ⌊4.5-7⌋. For $z = z_R$, the beam diameter is $2^{1/2}$ ($2w_0$). Equation (4.3) shows that with decreasing spot size, the accurate positioning of the substrate in the focal plane becomes more and more difficult. The total laser power is given by

$$P = 2\pi \int_0^\infty rI(r) \ dr = \pi w_0^2 \ I(0)/2 \ . \tag{4.4}$$

It should be noted that throughout this book, the laser power always refers to the *effective* power within the reaction chamber.

For some applications, homogenized or modified beam profiles are preferred, especially when structures with steep edges are required. For example, the beam may be homogenized by passing it through a curved quartz rod as described by CULLIS et al. ⌊4.8⌋. Structuring of the beam can be achieved by introducing into the beam path a metallic grid or an optical grating.

High power pulsed lasers are less desirable for direct writing of microstructures. With such lasers, a very high power is available for the short time, τ, of the pulse (Table 4.1). This may lead to a number of severe and fundamental problems: damaging of the substrate surface in laser-induced deposition or in synthesis from multilayered thin films; worse uniformity and morphology of generated patterns due to uncontrolled surface melting, ablation, or evaporation; worse localization and contrast of patterns due to homogeneous reactions within the adjacent reactive medium. For example, in gas-phase deposition, particulates may be homogeneously formed well below the dielectric breakdown. Such particulates may condense everywhere on the substrate surface or the surfaces of the reaction chamber, including the chamber windows. Besides these more fundamental problems, further problems arise with the presently available commercial pulsed lasers. These include the less effective control of the output power, at least on a shot to shot basis, and of the beam profile. Direct writing can be performed only with low scanning velocities, due to the low pulse repetition rates of visible and UV lasers.

There are, however, a number of applications where high power pulsed lasers are ideal sources. These are various kinds of surface modifications where rapid heating and cooling rates (see Sect.2.1.1 and Fig.2.3) and/or good energy localization within a thin surface layer are required. An example where these properties are a prior condition is the sheet doping of semiconductor surfaces (Sect.6.3). High power lasers in cw or pulsed operation are also desirable in large-area processing as described in Sects.4.1 (Figs.4.2b and 4.3) and 5.3. In this case, processing times can be significantly reduced by using a large-area focus or by initiating a homogeneous reaction in a large volume above the substrate surface. CO_2, Nd:YAG, Nd:glass and excimer lasers are most commonly used for this application.

High power lasers, and in particular excimer lasers, are gaining increasing interest for projection patterning. The preference for excimer lasers in this application is not only related to their high photon energy, which allows direct photodissociation of many molecules, but also to their relatively poor coherence (an excimer laser's output is highly multimode and contains as many as 10^5 transverse modes). The latter property avoids interference effects which cause severe problems in imaging applications with light of high spatial coherence.

4.3 Deposition and Etch Rates

In this section we will discuss the most commonly used techniques for measuring deposition or etch rates *in situ*. The practicability of a special technique depends on the size of these rates and on the particular experimental conditions and arrangements discussed in the preceding section.

In cw laser processing with a fixed beam and substrate, the beginning of deposition or etching is often indicated by the occurrence of a characteristic speckle pattern. With the onset of speckle movement, regular growth or etching starts. The observation of the speckles is a qualitative but very sensitive method, used mainly in gas-phase processing.

Sometimes, deposition rates can be measured from the transmission of the laser beam or a probe beam through the deposited film. This technique is primarily used in photolytic deposition where, because of the lower deposition rates, the thickness of films is small. Estimation of the film thickness requires knowledge of the optical constants of the substrate and the deposited film. For thin films, for example, of metals, these optical constants may differ considerably from the bulk values and, additionally, they may change with the film thickness. As a consequence, such estimates often involve tremendous inaccuracies. Transmission measurements are sometimes also used for measuring etch rates, especially of thin films or slabs.

Another very sensitive method of measuring deposition or etch rates over larger areas uses a microbalance (a piezoelectric slab-shaped crystal or ceramic). Here, the reaction yield, in terms of mass changes per unit time, can be directly derived from frequency changes of the microbalance. In laser-induced chemical etching, the material to be etched is first evaporated as a thin film, or glued as a thin platelet, onto the microbalance. The materials most commonly used for microbalances are crystalline quartz and ceramic PZT ($PbTi_{1-x}Zr_xO_3$). Fundamental investigations on quartz microbalances (QCM) have been performed in great detail $\lfloor 4.9$-$12\rfloor$. CHUANG and co-workers $\lfloor 4.13$-$15\rfloor$ have used quartz microbalances extensively to investigate laser-induced chemical etch rates. Sensitivities of about $1.23 \cdot 10^{-8}$ gr/Hz cm^2 have been obtained. For evaporated Si films, for example, this corresponds to about $2 \cdot 10^{14}$ atoms/Hz. In other words, an increase in frequency of 1 Hz due to the removal of less than one-half of a monolayer of Si can be readily detected. A typical frequency response of a QCM covered with a thin evaporated Si film to pulsed CO_2 laser irradiation is shown in Fig.4.5. In this experiment, the

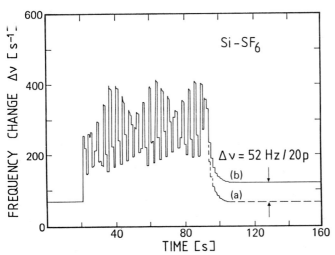

Fig.4.5. Frequency response of a quartz microbalance, covered with a Si film, under pulsed CO_2 laser irradiation (ϕ = 1 J/cm^2 at 942.4 cm^{-1}). (a) Vacuum; (b) $p(SF_6)$ = 2.7 mbar (after $\lfloor 4.15\rfloor$)

65

microbalance was positioned perpendicular to the unfocused laser beam. The momentary increase in microbalance frequency during each laser pulse is due to the temporary temperature rise caused by the absorbed laser light energy. In the absence of any reactive gas, the microbalance returned to its original frequency (curve a). Thus, under the experimental conditions used, the laser radiation by itself did not cause any significant removal of Si atoms from the solid surface. In a reactive gaseous atmosphere, e.g. in SF_6, an increase in microbalance frequency was observed. The frequency change of $\Delta\nu = 52$ Hz/20 pulses shown in curve b of Fig.4.5, corresponds to about $4.4 \cdot 10^{14}$ Si atoms removed per pulse. The experimental results of these investigations are further discussed in Chap.8. Different thermal and acoustic techniques for monitoring pulsed laser processing have been recently reviewed by MELCHER [4.16].

Deposition rates in the steady growth of rods (Sect.5.2.2) can be conveniently measured from the change in length of the rod, Δh, within a fixed time interval, Δt. This can be done by using a microscope, observing through a window of the reactor perpendicular to the laser beam, or automatically, by projecting an image of the hot tip of the rod onto the target of an optical multichannel analyser (OMA) or onto a simple position-sensing diode [4.17-19]. A similar configuration has been used to measure etch rates in laser-induced ablative photodecomposition (APD) of organic polymers [4.20]. Recently, the time evolution of APD has been investigated for various polymers by using wide bandwidth polyvinylidene-fluoride (PVDF) piezoelectric transducers to measure transient stress waves that are generated due to the recoil momentum transferred from the ablating material [4.21].

4.4 Temperature Measurements

One of the fundamental parameters in laser processing is the local laser-induced temperature rise. Therefore, a great variety of experimental techniques have been investigated to measure the temperature within the processed area [4.16-18,22-34]. Among these techniques are: visual and photoelectric pyrometry, pyroelectric calorimetry and photothermal deflection.

For applications in LCP, photoelectric pyrometry seems to be particularly promising. Here, the local temperature is derived from the optical radiation emitted from the laser-heated surface. The main advantages of this technique are the high temperature sensitivity, the relatively low sensitivity to surface properties and the suitability for in situ measurements. The spatial resolution is limited by diffraction to about one wavelength. The temperature can be derived from the emitted radiation in various ways: from its intensity at a certain wavelength, its intensity integrated over a broader spectral region or from its spectral dependence. In any case, this analysis requires additional information about the emissivity. This puts one of the main constraints on the accuracy of the technique. For most of the materials investigated in LCP, only little is known about the magnitude, the spectral dependence and the temperature dependence of the emissivity. Furthermore, the emissivity of a material in thermal equilibrium can differ considerably from the emissivity of the same material under laser irradiation. The reason for this difference is based on the localization of the laser-induced temperature distribution perpendicular to the surface (z-direction) and, in the case of semiconductors, on the high concentration of photogenerated nonequilibrium carriers. The effect of localization is henceforth described by an apparent emissivity ε_{app}, and the emissivity of the same material at a homogeneous temperature by ε. Figure 4.6 shows the ratio of ε_{app} to ε, plotted as a function of photon energy $\lfloor 4.32 \rfloor$. The temperature variation in the z-direction was approximated by $T(z) = (T_0 - 300)w/(w + z) + 300$ with $w = w_0/2^{1/2}$ (Sect.2.1.1). The optical absorption coefficient α was assumed to be constant. The figure clearly shows that even for values $\alpha \cdot w = 10$, ε_{app} and ε may differ considerably. In describing actual experimental situations, the

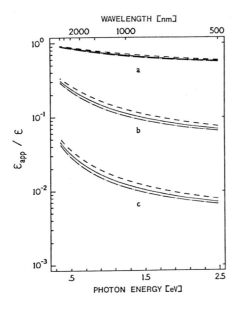

Fig.4.6. Ratio of the apparent emissivity due to localization, ε_{app}, to the emissivity of the same material at homogeneous temperature, ε, as a function of the photon energy. The surface temperatures (z = 0) are 1000 K (dash-dotted curves), 1200 K (full curves) and 1500 K (dashed curves). The optical absorption coefficient α is assumed to be constant. The parameters $w \cdot \alpha$ are: (a) 10, (b) 1, (c) 0.1 (after $\lfloor 4.32 \rfloor$)

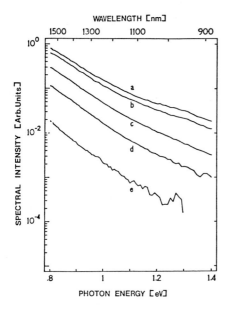

WAVELENGTH [nm]

SPECTRAL INTENSITY [Arb.Units]

PHOTON ENERGY [eV]

Fig.4.7. Spectra of the radiation emitted from the laser-heated tip of a Si rod (length 5 mm, diameter 100 μm) grown by LCVD from SiH_4. (a) P = 150 mW (1360K), (b) 140 mW (1300K), (c) 120 mW (1150K), (d) 100 mW (1080 K), (e) 80 mW (900 K). The temperature values given in parentheses are derived from fits to Planck's law (after ⌊4.32⌋)

temperature dependence of α must be taken into account. In Si, for example, the increase in carrier density due to laser excitation causes an increase in absorption and thereby an increase in the apparent emissivity (Sect.2.1.1). Measurements of the thermal radiation emitted from various laser-heated materials, and especially from semiconductors, have been reported in a number of papers ⌊4.17,27-34⌋. In some of these experiments, however, $\alpha \cdot w$ was not large enough to allow the difference between ε_{app} and ε to be neglected. In these cases, the temperatures derived from the intensities or the spectra of the emitted radiation are erroneous. Reliable in situ measurements in LCP were performed during steady growth of C and Si rods (see Sect.5.2.2 and ⌊4.17,18,32,34⌋). These experiments employed visual and photoelectrical micropyrometry. The melting point of Si was reproduced to within ± 10K. Typical spectra of the radiation emitted from laser-heated tips of Si rods are shown in Fig.4.7. These spectra can be fitted well with Planck's law. The temperatures derived from the fit are in good agreement with temperatures determined by visual pyrometry ⌊4.17,34⌋. One of the reasons for the good agreement, however, is based on the fact that the temperature gradients in quasi-one-dimensional structures such as rods are much less pronounced than in semiinfinite substrates. Therefore large values of $\alpha \cdot w$ can be realized more easily.

5. Material Deposition

Laser-induced materials deposition has been performed from adsorbed, gas, liquid and solid phases. These various different possibilities have been shown schematically in Fig.3.1. Table 5.1 summarizes data on the elements and compounds that have so far been deposited. Additional information on the physical and chemical properties of oxides and nitrides deposited from the gas phase is listed in Tables 5.2-4. Surface modifications such as surface oxidation and nitridation are treated in Chap.6 and the systems investigated are listed separately in Table 6.1. This holds also for surface doping (Table 6.2), while the deposition of doped films is included in Table 5.1. Compounds that are formed from multilayered structures or powders are also not included in Table 5.1, but are cited only in Chap.7. It should be emphasized that for some of the systems listed, the results should be taken as preliminary. This limitation applies not only to the deposition rates but also to the purity and stoichiometry of the materials. A brief glance at Table 5.1 shows that most of the investigations have concentrated on laser-induced deposition from the gas phase, henceforth abbreviated as LCVD. Liquid-phase deposition, and in particular laser-enhanced electrochemical deposition, has been demonstrated mainly for metals. Table 5.1 also contains some results on thin film formation by photochemical processing with lamps, including photosensitized reactions, and by laser sputtering and evaporation. Laser-induced formation of organic layers has been demonstrated only for polymethyl methacrylate (PMMA) by photo-polymerization of gaseous MMA.

This chapter is organized in the following way. In Sect.5.1 we make some general remarks on the initial phase of growth, nucleation. Then, in Sects.5.2 and 5.3, we discuss the deposition of materials in the form of microstructures and extended thin films, respectively. This discussion concentrates mainly on model systems that have been investigated in great detail. However, this discussion is very general in the sense that most of the trends, features and results apply to all the other systems listed in the table but not discussed explicitly throughout this chapter.

Table 5.1. Inorganic materials deposited from the gas, liquid or solid phase. Processing is performed at perpendicular cw laser incidence unless otherwise specified. Pulsed (p-) laser irradiation is only indicated in cases where cw irradiation would also be possible. The corresponding laser light intensities I (cw processing) and fluences ϕ (pulsed processing) in W/cm² and J/cm² respectively, are given in parentheses. The abbreviation XEY denotes X · 10^Y. Laser wavelengths are rounded and given in nm (except those of CO₂ lasers, which are given in μm). The deposition rates, W, are in most cases the highest rates presently achieved for controlled processing. d and h are the width and height of laser-processed structures, respectively. Most of the abbreviations employed are listed at the end of the book. The following additional contractions have been used: S deposition of spots; L direct writing; R growth of rods

Solid	Parent/Carrier	Substrate (h_L [μm])	Laser λ [nm, μm]	W [μm/s] (I [W/cm²]; ϕ [J/cm²]) d [μm], h [μm]	Remarks	Ref.
Ag	AgCF₃SO₃[a], AgPF₆, AgBF₄/C₆H₅CH₃, CH₃CN sol.	Glass[b], SiO₂	325 HeCd	W = 1E−4 (1E−3) d = 2.5	L, ρ, SIMS; projection patterning	[5.1]
Ag	AgNO₃/N-methyl-2-pyrrolidinone sol.	Ceramic Al₂O₃	Ar		L, $\rho_D \approx \rho_B$, SEM, X	[5.2]
	Spun-on AgNO₃ in polymer host	Glass	Ar	d = 10	L, $\rho_D \approx \rho_B$, UV, VIS	[5.3]
Al	Al₂(CH₃)₆	c-Al₂O₃	ArF, KrF	(3E−4)	S, AL, HS 400, P, λ, p, PP, IR; optoacoustic	[5.4]
Al	Al₂(CH₃)₆	(100) Si	KrF		p, UV, VIS; photofragment analysis	[5.3]
Al	Al₂(CH₃)₆/H₂	Glass, SiO₂, Si[c], GaAs, InP	ArF, KrF	W = 1.7E−3	⊥, ∥, EF, HS 500, $\rho_D \approx 1-3\ \rho_B$	[5.5, 6]
Al	Al₂(CH₃)₆	SiO₂, c-Al₂O₃, Si	150 VUV lamp (Ar, Kr, Xe)		λ, F	[5.7]
Al	Al₂(CH₃)₆/He	Glass, SiO₂	ArF, 257 Ar	$W \leq$ 1E−4	S; ripples, on prenucleated Zn	[5.8–13]
Al	Al₂(CH₃)₆	c-SiO₂	257 Ar		AL, P, T, UV, VIS; adsorption isotherms	[5.14]
Al	Al₂(CH₃)₆	SiO₂, Si	257 Ar		AL, p, T; ripples	[5.15–17]

Al	Al$_2$(CH$_3$)$_6$/H$_2$	(100) Si	ML Ar + 254 Hg UV lamp	$W \leq 10$	T; spikes, a-Al, Al$_3$C$_4$	[5.18]
Al	Al$_2$(CH$_3$)$_6$/H$_2$	(100) GaAs	476–647 Kr		CO$_2$ laser CVD on prenuculated L	[5.19, 20]
Al	AlI(C$_4$H$_9$)$_3$[d]	SiO$_2$	257 Ar			[5.16, 21]
Al	AlI$_3$	Stainless steel, Ni, Cu	ArF	$W = 1E-5$, $h < 0.1$	F	[5.22]
Al$_x$Ti$_y$	Al$_2$(CH$_3$)$_6$ + TiCl$_4$		257 Ar		AL, p, K; C$_2$H$_2$, C$_2$H$_4$ polymerization	[5.16, 17, 23, 24]
Al$_2$O$_3$	Al$_2$(CH$_3$)$_6$ + N$_2$O/Ar	Si	ArF, KrF		$\perp, \parallel$, EF, HS 700, $\rho_D \leq 10^{16}$	[5.25–27]
Al$_2$O$_3$	Al$_2$(CH$_3$)$_6$ + N$_2$O/He	Glass, Si, GaAs, InP	ArF, KrF	$W = 0.3$	$\perp, \parallel$, EF, HS 300–700, $\rho_D = 1E11$	[5.5, 6, 28, 29]
Au	Au(CH$_3$)$_2$(AcAc)[e]	0.3 SiO$_2$/Si, Si	515 Ar	$W = 1$ (3E5)	S, L, $\rho_D \approx 4\rho_B$	[5.30, 31]
Au	Gold-cyanide[f] sulfite sol.	1 Ni/Be-Cu, Ni	MLAr	$W = 1$ (2E4)	S, L, HS 320, P, PP, SEM; jet-LEP,	[5.32–36, 222]
Au	HAuCl$_4$/aq.	(100) Si	532 p-Nd:YAG	$W \leq 30$	S, L, SEM, RBS	[5.37]
Au	Au-salt sol.	0.1 Mo, Ni, W/glass;	647 Kr	$W = 1$	S, L, LEP; electroless plating	[5.32, 33, 38–41]
Au	HAuCl$_4$/CH$_3$OH aq.	0.1 Mo, Ni, W/c-Al$_2$O$_3$	580–720 p-dye/XeCl	$h = 0.03$ (0.2)	SEM; ohmic contacts	[5.42]
Au	Au-cyanide/aq.	(100)InP	W lamp (IR cutoff)		Electroplating, low adhesion	[5.43]
Au	Aurotron[g]	Si, GaAs	515 Ar	$W = 6E-4$ (200)	Electroless plating	[5.44]
Au	Spun-on metallopolymer[h], $h = 1.5$ (18 wt.% Au)	Cu/glass	515 Ar	$d \geq 3$, $h = 0.17$ (1E3)	S, L, P, $\rho_D \approx 10 \rho_B$, K, AES, RBS; periodic structure	[5.45–47]
C	C$_2$H$_2$, C$_2$H$_4$, CH$_4$	SiO$_2$	488 Ar, 647 Kr	$W \leq 100$ ($\leq$3E5)	S, R, L, P, λ, p, T, K, SEM	[5.48–54]
C	CH$_4$ + C$_2$H$_6$ (natural gas)	Ceramic Al$_2$O$_3$, C, W	CO$_2$	$W = 8$(3E4)	R, X	[5.55]
C	C$_2$H$_5$Cl/Ar, N$_2$	NaCl, KBr, Si	CO$_2$	$W = 1E-3$	$\perp, \parallel$, EF, ρ, K, IR, RA	[5.56]

Solid	Parent/Carrier	Substrate (h_L [μm])	Laser λ [nm, μm]	W [μm/s] (I [W/cm²]; ϕ [J/cm²]) d [μm], h [μm]	Remarks	Ref.
Cd	Cd(CH₃)₂	Glass, SiO₂, C	ArF, 257 Ar	—, $W = 0.1(1E4)$	S, L, AL, p, K, UV, SEM, TEM, AES; adsorption isotherms prenucleation techn.	[5.8–11, 13, 14, 57]
Cd	Cd(CH₃)₂	Glass, C	257 Ar	$W \leqslant 0.01$	S, SEM, TEM; ripples, angular dependence	[5.12]
Cd	Cd(CH₃)₂	SiO₂	257 Ar	$W \leqslant 1E-3$	AL, ripples	[5.15]
Cd	Cd(CH₃)₂	SiO₂	257 Ar		AL vs. gas-phase processes	[5.58, 59]
Cd	Cd(CH₃)₂/Ar	Cd spheres on C films	257 Ar		Local electric field EM	[5.60, 61]
Cd	Cd(CH₃)₂/Ar	SiO₂	257 Ar	$W = 5E-3$	AL; gratings	[5.62]
Cd	Cd(CH₃)₂/Ar	Plastic SiO₂/Si, Si	257 Ar		L, P, p, ρ, UV, VIS, SEM	[5.63]
Cd	Cd(CH₃)₂	c-SiO₂	254 Hg lamp		P, p	[5.64]
Cd	Cd(CH₃)₂/H₂	(100) GaAS, SiO₂	337–676 Kr	$W = 0.02$ (1.3E5)	S, L, AL, P, λ, T, UV, VIS, SEM; luminescence, latent time	[5.19, 20, 65]
Cd	CdSO₄ + H₂SO₄/aq.	Si	W lamp		Electroless plating	[5.66]
CoO	Co(AcAc)₃ °/N₂	Glass	CO₂		P, T, SEM; model calculations	[5.67]
Cr	Cr(CO)₆/He	Glass, SiO₂, SiO₂/Si, Si	ArF, KrF, 260–270 Cu(II)	$W = 3E-3$ ($\perp$) – ($\perp$)	$\perp$, ‖, S, EF, HS 420, PP, $\rho_D \approx 16 \rho_B$, SEM	[5.6, 68, 69]
Cr	Cr(CO)₆	Glass, SiO₂, Al	ArF, KrF, XeF	(12E-3)	$\perp$, ‖, SEM, AES, X	[5.199]

72

Element	Precursor	Substrate	Light source	W (d, h)	Characterization / Notes	Ref.
Cr	Cr(CO)$_6$/Ar	SiO$_2$	KrF; 257, 515 Ar		S, L, P, λ, ρ, SEM, AES; projection	[5.70]
Cr	Cr(CO)$_6$/He, Ar	Glass, SiO$_2$, Si	257 Ar	$W = 1\mathrm{E}{-3}$	L	[5.71]
Cr	Cr(CO)$_6$	SiO$_2$, Si	280–350 dye/ 532 Nd:YAG	$W = 2\mathrm{E}{-4}$	S, L, AL, P, λ, SEM, AES	[5.72]
Cr	Cr(C$_6$H$_6$)$_2$/C$_6$H$_6$	SiO$_2$	488 Ar	$W = 20$ (6E6) $d \approx 15$ $h \approx 1$	S, L, P, VIS, SEM, AES; $C \approx 10\%$	[5.228]
Cr$_2$O$_3$/CrO$_2$	CrO$_2$Cl$_2$	Glass, SiO$_2$, (100) Si, GaAs	488–515 Ar	$W = 3$ (5E6)	R, L, p, AES, XPS, X; single crystals	[5.73]
Cu	Cu(HFAcAc)$_2$^{i}	0.002–0.1 SiO$_2$/Si, Si	515 ML Ar	$W = 270\ \mu m^3/s$ (5E5)	S, L, EF, P, $\rho_D \approx 2\,\rho_B$, SEM	[5.74, 75]
Cu	Cu(HFAcAc)$_2$^{i}	Glass	515 Ar	$d < 10$ (6E4)	L, SEM, $\rho_D \approx 3\,\rho_B$	[5.76]
Cu	Cu(HFAcAc)$_2$^{i}/ C$_2$H$_5$OH	c-SiO$_2$, Si	ArF, KrF, 257 Ar, Hg arc lamp	— $W \leqslant 3\mathrm{E}{-4}$ (1E4) —	S, AES, XPS	[5.77]
Cu	CuSO$_4$ + H$_2$SO$_4$ aq.	Ni, Be$_x$Cu$_y$, stainless steel	Ar	$W = 50$ (6E3)	S, P, $\rho_D \approx \rho_B$, SEM; jet-LEP	[5.32, 33, 78]
Cu	CuSO$_4$ + H$_2$SO$_4$ aq.	0.05–1 Au, Cu, Mo, W/glass	ML Ar, Kr	$W \leqslant 5$ (2E3) $d = 2$	EM 500, LEP, SEM; electroless plating	[5.32, 33, 38, 39, 41]
Cu	CuSO$_4$ + H$_2$SO$_4$ + HCl/aq.	Cu/glass; 0.02 Ni/glass, Cu/phenolic resin paper	515 Ar	$W \leqslant 0.07$ (200) $W = 0.1$	S, L, SEM	[5.44]
Cu	Cu(HCOO)$_2$		1064 Nd:YAG, Nd:glass, CO$_2$	(1.4E3)	ρ	[5.79]
Cu	Cu$_2$P$_2$O$_7$/aq. CuClO$_4$/CH$_3$CN	GaAs, Si	W lamp (IR cut off)	$W = 7\mathrm{E}{-4}$	Electroplating, low adhesion	[5.43]
Fe	Fe(CO)$_5$	Al$_2$O$_3$, (100) GaAs	ArF, KrF	$W = 2\mathrm{E}{-5}$ $h < 0.05$	EF, PP, $\rho_D \approx 20$–$30\,\rho_B$, SEM, AES, X	[5.80]
Fe	Fe(CO)$_5$	GaAs	Hg lamp		AL, UV, VIS	[5.81]
Fe	Fe(CO)$_5$	(100) Si	Hg lamp		Al, HS 270–870, AES; EBCVD	[5.82]

73

Table 5.1 (continued)

Solid	Parent/Carrier	Substrate (h_L [μm])	Laser λ [nm, μm]	W [μm/s] (I [W/cm²]; ϕ [J/cm²]) d [μm], h [μm]	Remarks	Ref.
Fe	Fe(CO)₅/He, Ar	Glass, SiO₂, Si	257 Ar	$W = 3E-3$ (600)	L, P	[5.71]
Fe	Fe(CO)₅	SiO₂	ML Ar		HS 500, SEM, AES; periodic structure due to phase segregation	[5.83]
Fe	Fe(CO)₅	SiO₂	p-CO₂	$W = 10$ (50)	S, PP	[5.84, 85]
Ga	Ga(CH₃)₃/H₂	SiO₂	ArF 257 Ar 257 mode-locked	– $W = 1.7E-3$ $W = 0.1$	S, UV	[5.20, 86]
Ga	Ga(CH₃)₃	Glass, (100) GaAs, (100) InP	Xe arc lamp		SEM, AES	[5.87]
GaAs	Ga(CH₃)₃ + AsH₃/H₂	(100) GaAs	532 p-Nd:YAG	$W = 1E-3$	HS 720, SEM; enhanced MOCVD, epitaxial	[5.88, 89]
GaAs	Ga(CH₃)₃ + AsH₃/H₂	(100) GaAs	458–515 Ar	(1E3)	S, HS 800–950, enhanced MOCVD	[5.90]
GaAs	Ga(CH₃)₃ + AsH₃/H₂	(100) Si	458–515 Ar	$d = 40$ $h = 1.7$	S, L, HS 300–720, PL	[5.91]
GaAs	Cl(CH₃)₂ Ga, As(C₂H₅)₃/H₂, Cl(CH₃)₂ Ga, As(CH₃)₃/H₂	GaAs Ge	647 Kr	$W < 0.01$		[5.92]
Ge	GeH₄/He	SiO₂ (100) NaCl (1̄102) Al₂O₃	ArF, KrF, XeCl	$W = 0.3$ Å/p	HS 800, P, λ, PP, ρ, VIS, TEM, X; epitaxial film on NaCl, HS 400	[5.93–96]
Ge:Al	GeH₄ + Al₂(CH₃)₆/He		ArF		AES, 10^{20}/cm³ Al	[5.94]

HgTe	$Te(C_2H_5)_2/H_2$ + Hg vapor	(100) InSb	Hg lamp	$W = 6E-5$	EF, HS 600, SEM, X; photosensitization, epitaxial	[5.97, 98]
$Hg_{1-x}Cd_xTe$	$Hg(CH_3)_2$ $+Cd(CH_3)_2$ $+Te(CH_3)_2/He$	CdTe	ArF	$W = 1E-3$	‖, HS 420, IR, SIMS, X; epitaxial	[5.229]
In	InI	Stainless steel, Ni, Cu	ArF	$W = 1.3E-5$ <0.1 μm	F, AES	[5.22]
Ir	$In(CH_3)_3$, InC_5H_5, $(CH_3)_3InP(CH_3)_3$[j], $(CH_3)_3InP(C_2H_5)_3$	Pyrex[b], (100) GaAs, 100 InP	Xe lamp		SEM, AES; C islands, whiskers of InP	[5.87]
InP	$(CH_3)_3InP(CH_3)_3$[j] $+ P(CH_3)_3/He + H_2$	SiO_2, GaAs, InGaAs, (100) InP, Si	ArF KrF XeF	$W = 0.4$ Å/p (0.04) $h = 1.15$ –	EF, HS 600, K, UV, VIS, IR, SEM, TEM, AES, RBS; amorphous, epitaxial	[5.99–101]
In_2O_3	$(CH_3)_3InP(CH_3)_3$[j] $+ P(CH_3)_3 + O_2$ or H_2O vapor/H_2, He	SiO_2, GaAs, InP	ArF	$W = 0.2$ Å/p	⊥, ‖, HS 600, n, K, UV, VIS, AES	[5.100, 101]
Mn	$Mn_2(CO)_{10}$	Si, ZnS	337–356 Kr		L, AES; thermal indiffusion electroluminescent devices	[5.102]
Mo	$Mo(CO)_6$	Glass, SiO_2, Si	ArF, KrF	$W = 4E-3$ (⊥)	⊥, ‖, EF, HS 420, PP, $\rho_D \approx 7\rho_B$	[5.6]
Mo	$Mo(CO)_6$	Glass, SiO_2, Al	ArF, KrF, XeF	(12E-3)	⊥, ‖, SEM, AES, X	[5.199]
Mo	$Mo(CO)_6/He$	Glass, SiO_2, SiO_2/Si, Si	ArF, KrF, 260–270 CuI	$W = 4E-3$ –	S, EF, PP, $\rho_D \approx 20\,\rho_B$	[5.68, 69]
Mo	$Mo(CO)_6$	Glass, Al_2O_3, GaAs	350–360 Ar	$W = 2.7E-3$	L, HS 360, P, ρ	[5.103]
Mo	$Mo(C_6H_6)_2/C_6H_6$	Pyrex	488 Ar	$W = 7$ (9E6)	S, P, VIS, SEM, AES; C ≈ 10%	[5.228]
Ni	$Ni(CO)_4/He$	Glass, 0.1 a-Si/glass, 0.4 μm SiO_2/Si	476–647 Kr	$W \leqslant 100$ (5E5) $W = 10$ (2E6)	S, R, L, AL, P, λ, p, PP, $\rho_D \approx 2–3\,\rho_B$, K, SEM	[5.48, 50, 104, 105]

Table 5.1 (continued)

Solid	Parent/Carrier	Substrate (h_L [μm])	Laser λ [nm, μm]	W [μm/s] (I [W/cm²]; φ [J/cm²]) d [μm], h [μm]	Remarks	Ref.
Ni	Ni(CO)₄	0.1 a-Si/glass	647 Kr		Periodic structures	[5.106]
Ni	Ni(CO)₄/He	Si	515 Ar	$W = 100$ (3E6)	L, P, K; model calculations	[5.107]
Ni	Ni(CO)₄	a-Si/glass	647 Kr	Lateral growth rate 300 (1E5)	S, P, λ, p, T, K, SEM; model calculations	[5.48, 108]
Ni	Ni(CO)₄	SiO₂	10.6 CO₂	$W = 15$ (225)	S, P, PP, $\rho_D \approx 3–10\,\rho_B$, SEM	[5.85, 109–112, 189]
Ni	Ni(CO)₄/Ar	Glass	10.6 p-CO₂	$W = 8$ Å/p	‖, PP, AES, X; $C \approx 10\%$	[5.230]
Ni	NiCl₂/aq.	0.1 Mo, Ni, W/glass, /c-Al₂O₃, /metals	ML Ar	$W \leq 5$ (5E3) $d \geq 4$ $h \leq 0.4$	EM 10³, LEP; electroless plating	[5.32, 33, 41]
Ni	NiSO₄/aq.	(100) InP	580–720 p-dye/XeCl	$h \leq 0.5$	Photoelectrochemical, low adhesion	[5.42]
Ni	Ni(NH₂SO₃)₂-bathᵏ/aq.	GaAs, Si	W lamp			[5.43]
Pb	Pb(C₂H₅)₄	c-SiO₂	254 Hg lamp		AL, P, p	[5.114]
Pd	Palladium cyanide/aq.	GaAs, Si	W lamp		Photoelectrochemical	[5.43]
Pd	[Pd(μ-O₂CCH₃)₂]₃ (spun-on)	SiO₂, Si	515, ML Ar		L; in air PdO	[5.115]
Pt	Pt(PF₃)₄	Al, C, Cu, Si, SiO₂	KrF, XeCl	$W = 3$ Å/p (0.2)	S, K, SEM, AES, RBS	[5.116–118]
Pt	Pt(HFAcAc)₂ⁱ	a-C	257, 515 Ar		S, TEM	[5.119]
Pt	Pt(HFAcAc)₂ⁱ	Glass, Al₂O₃, GaAs	350–360 Ar	$W = 2E-5$	L, HS 360–420, P, p, ρ	[5.103]
Pt	H₂PtCl₆, CH₃OH/aq.	(100) GaAs, (110), (111) InP	580–720 p-dye/XeCl	$h \leq 0.5$	SEM, TEM, AES; ohmic contacts, no dep. on Si	[5.42]

Material	Precursor	Substrate	Laser/Source	W	Methods	Ref.
Pt	Spun-on metallo-polymer[h] (7.5 wt.% Pt)	Glass	XeCl 488, 515 Ar	$d = 5, h = 0.1$	P, ρ	[5.221]
Se	Se(CH$_3$)$_2$/Ar	Al$_2$O$_3$, SiO$_2$	Hg/Xe lamp	$W = 1E-4$	EF, SEM	[5.120]
Si	SiH$_4$/N$_2$, (Al$_2$(CH$_3$)$_6$)	SiO$_2$	ArF, KrF, XeCl	$W = 5E-4$	HS 400, PP, ρ, TEM, X; Al doping	[5.93, 94]
Si	SiH$_4$	SiO$_2$, (100) Si	KrF	$W = 2E-3$ $W = 6E-4$	RA, TEM; poly- and epitaxial LPCVD	[5.121]
Si	SiH$_2$Cl$_2$/H$_2$	(100) Si	Hg-Xe lamp		EF, HS 1000, ρ, RA; epitaxial	[5.122]
Si	SiH$_4$	(100), (110), (111) Si	488 Ar, 531 Kr	$W = 30$	R, P, λ, T, PP, K, RA, SEM, X; single cryst.	[5.48, 50, 104, 123, 124]
Si	SiH$_4$	Glass, a-Si/glass	488 Ar	$W \leqslant 40 \ (\leqslant 4E5)$	R, L, P, λ, T, K, SEM	[5.125, 126]
Si	SiH$_4$	a-Si/glass, Si	515 Ar, 647 Kr		Periodic structures	[5.125, 127, 128]
Si	SiH$_4$ + BCl$_3$, B(CH$_3$)$_3$, Al$_2$(CH$_3$)$_6$	SiO$_2$/Si, Si	ML Ar	$W = 15 \ (2E7)$	L, P, p, PP, ρ, R, SEM, X; B, Al doping	[5.57, 129]
Si	SiH$_4$ + B$_2$H$_6$, PH$_3$	SiO$_2$/Si, Si	515 Ar		L, ρ, RA, SIMS; B, P doping, nucleation model	[5.130−134]
Si	SiH$_4$	Glass, Al$_2$O$_3$, SiO$_2$, C	10.3, 10.6 CO$_2$	$W = 0.02$	$\perp, \parallel$, HS 800, K; SF$_6$ sensitizer; in $\parallel$ probably a-Si:H	[5.135]
Si	SiH$_4$/H$_2$, Ar	SiO$_2$	10.3, 10.6 CO$_2$	$W = 0.1$	S	[5.136]
Si	SiH$_4$	Glass, SiO$_2$	10.6 p-CO$_2$		λ, p	[5.137]
Si	SiH$_4$	SiO$_2$	CO$_2$, 10.6 p-CO$_2$	$W = 1.5$ Å/p ($\perp$) $W = 0.25$ Å/p ($\parallel$)	$\perp, \parallel$, S, EF, HS 300−700, P, p, RA; in $\parallel$ a-Si:H	[5.138−140]
Si	SiCl$_4$/H$_2$	SiO$_2$	10.5 CO$_2$	$W = 0.1$	S, P, K	[5.141]
a-Si:H	SiH$_4$, Si$_2$H$_6$	SiO$_2$	ArF, KrF		p, VIS, UV	[5.142]
a-Si:H	Si$_2$H$_6$/He	SiO$_2$	ArF, 254 Hg lamp	$W = 1E-4 \ (\parallel)$ $W = 1E-4$	EF, HS 400−700, p, ρ, IR; hybrid LCVD + PCVD, P doping, Hg photosensitization	[5.143, 144]

Table 5.1 (continued)

Solid	Parent/Carrier	Substrate (h_L [µm])	Laser λ [nm, µm]	W [µm/s] (I [W/cm^2]; ϕ [J/cm^2], d [µm], h [µm])	Remarks	Ref.
a-Si:H	Si_2H_6, Si_3H_8/He, Ar	Glass, Si	185 Hg lamp	(3.5E−3)	HS 600, p, ρ	[5.227]
a-Si:H	Si_2H_6	SiO_2	254 Hg lamp	$W \leqslant 2$E−3	EF, HS 500−600, p, ρ; B, P doping	[5.145]
a-Si:H	Si_2H_6	Si	185, 254 Hg lamp	$W = 2$E−3	EF, HS 520, p, ρ, UV, VIS, IR	[5.146]
a-Si:H	SiH_4	Si	Ar+CO_2		S, L, SEM, IR, AES; twin-beam	[5.147]
a-Si:H	SiH_4, Si_2H_6	SiO_2	10.6 CO_2	$W = 0.2$ (10)	⊥, ∥, P, λ, p, ρ, X; CARS	[5.148−150]
a-Si:H	SiH_4/Ar, N_2 $SiH_4 + B_2H_6$, PH_3	Glass, SiO_2, C, Si	10.6 CO_2	$W = 1.5$E−3 (⊥) $W = 5$E−4 (∥) (700)	⊥, ∥, EF, HS 500−700, P, ρ, IR, X; B, P doping a-Si, C:H alloys	[5.135, 151−155]
a-Si:H	SiH_4/H_2, He, Ar	Si	10.6 CO_2	$W = 2.5$E−4	∥, EF, HS 500−700, ρ, UV, VIS, IR, ESR; B, P doping	[5.156−158]
a-Si:H	$RSiH_3$ (R = C_2H_5, C_4H_9, C_6H_5, SiH_3)	Ag, Mo	CO_2		SEM	[5.159]
a-Si:H	SiH_4, SiH_2Cl_2/N_2	Glass, SiO_2	10.6 p-CO_2	$W = 30$ Å/p (2) (⊥)	⊥, ∥, HS 600−700, P, p, IR; homogeneous mechanisms	[5.139, 140, 160, 161]
a-Si:H	H_2, SF_4, Si target	Glass	532, 1064 p-Nd:YAG	$W = 10$ (2)	p, ρ, UV, VIS, IR, TEM; reactive sputtering	[5.148, 162, 163]

Material	Reactants	Substrate	Laser	Parameters	Methods	Ref.
a-Si, C:H	$SiH_4 + C_2H_4$, C_3H_6		10.6 CO_2		p, UV, VIS; B, P doping	[5.155]
SiO_2	$SiH_4 + N_2O$	Si	ArF	$W = 1.5E-3$ ($\perp$, $\parallel$)	$\perp$, $\parallel$, EF, HS 400–800, n, PP, IR; combined $\parallel$, $\perp$ incidence	[5.164]
SiO_2	$SiH_4 + N_2O/N_2$	Si	ArF	$W = 3E-4$	$\parallel$, EF, HS 500, P, p; model calculations	[5.165]
SiO_2	$SiH_4 + N_2O/He$ $SiH_4 + N_2O/N_2$ $SiH_4 + N_2O/N_2$	Si (100) Si p-Si/SiO_2/Si	ArF	$W = 1E-3$ ($\parallel$) $W = 1.4E-3$ ($\parallel$) $W \leqslant 5E-3$ ($\parallel$)	$\perp$, $\parallel$, EF, HS 300–900, p, n, PP, K, IR, AES, SIMS; 9 MV/cm, good step coverage	[5.6, 28, 29, 166–168]
SiO_2	$SiH_4 + O_2 + N_2/H_2$	Si	KrF	$W = 5E-4$ (0.1)	$\parallel$, EF, HS 450–520, P, p, IR, AES	[5.169]
SiO_x	$SiH_4 + N_2O$	(100) Si	531 Kr	$W \leqslant 2$ (1E5)	S, R, L, P, p, n, RA, SEM; $x = 2$; $x < 2$	[5.123, 170]
SiO_2	Spun-on $Si(OR)_x(OH)_{4-x}$	SiO_2, CdS, GaAs, Si	515 Ar	$h = 0.5$ (4E6)	L, P, IR, SEM	[5.171]
Si_3N_4	$SiH_4 + NH_3/Ar$	SiO_2/Si, ZnSe, Si	ArF	$W = 1E-4$	$\parallel$, HS 500–900, P, n, AES	[5.25]
Si_3N_4	$SiH_4 + NH_3/N_2$	Si	ArF	$W = 1.2E-3$	$\perp$, $\parallel$, HS 320–700, PP, IR, ESCA	[5.6, 28, 29]
Si_xN_y	$SiH_4 + NH_3/N_2$	Si	CO_2	$W = 2E-4$	$\parallel$, EF, HS 500–720	[5.172]
Sn	$Sn(C_2H_5)_4$	InP	ArF, KrF		L	[5.173]
Sn	$Sn(CH_3)_4$	a-C	257, 515 Ar		S, TEM	[5.76, 119]
SnO_2	$(CH_3)_2SnCl_2 + O_2$	SiO_2	10.6 CO_2		AL, HS, p, T, SEM, X; autocatalysis	[5.174]
Ti	$TiCl_4/H_2$	SiO_2	257 Ar	$W = 0.1$ (<5E4) $d = 4$	AL, HS, p, T, SEM, X; autocatalysis	[5.16, 24]
Ti	$TiCl_4/H_2$, Ar	$LiNbO_3$, SiO_2	257 Ar		P, p; waveguides in $LiNbO_3$	[5.175]
TiC	$TiCl_4 + CH_4$	SiO_2, steel	10.6 CO_2			[5.110, 112, 176]
TiO_2	$TiCl_4 + H_2 + CO_2$	SiO_2	10.6 CO_2			[5.109, 110, 112]

Table 5.1 (continued)

Solid	Parent/Carrier	Substrate (h_L [μm])	Laser λ [nm, μm]	W [μm/s] (I [W/cm²]; φ [J/cm²]; d [μm], h [μm])	Remarks	Ref.
TiSi₂	TiCl₄ + SiH₄	Si	ArF CO₂	—	∥, HS 700, ρ, K, SEM, AES, RBS, X	[5.177, 178]
Tl	TlI, TlBr	Ni, Cu, stainless steel	ArF	$W = 3E-4$	EF, F	[5.22]
W	WF₆ + H₂	Al₂O₃, SiO₂, Si	ArF	$W = 1E-5$ $W = 5E-5$	HS 800, ρ, SEM, AES, X	[5.179]
W	WF₆/Ar	SiO₂/Si, Si	ArF	$W > 2E-3$	⊥, ∥, EF, HS 300–700, T, $\rho_D \approx 2\rho_B$, SEM, AES, X; good step coverage	[5.26, 180]
W	WF₆	Si	515 Ar		L, SEM, AES	[5.132]
W	WF₆	Si	ML Kr		S, L, SEM; small single cryst.	[5.85, 181]
W	WF₆	SiO₂	10.6 CO₂	(1.3E6)	ρ	[5.182]
W	WF₆	SiO₂	10.6 CO₂	$W = 10$	PP, SEM, $\rho_D \approx 3-10\,\rho_B$	[5.84, 85]
W	WF₆/Ar	Si	Ar		L, RA, AES, RBS, X; silicide formation	[5.183]
W	W(CO)₆/He, Ar	Pyrex[b], SiO₂, Si	ArF, KrF, 257 Ar	$W = 3E-4(1E3)$	L	[5.71]
W	W(CO)₆/He	Glass, SiO₂, SiO₂/Si, Si	ArF, KrF, 260–270 Cu II	$W = 3E-3$ (⊥)	⊥, ∥, S, EF, HS 420, PP, $\rho_D \approx 20\,\rho_B$	[5.6, 68, 69]
W	W(CO)₆	Glass, SiO₂, Al	ArF, KrF, XeF	(12E-3)	⊥, ∥, HS 310, SEM, AES, X	[5.199]
W	W(CO)₆	Glass, Al₂O₃, GaAs	350–360 Ar	0.035 — 0.018	L, HS 360, P, ρ	[5.103]

Material	Precursor	Substrate	Laser	Parameter	Techniques	Ref.
WSi_2	$WF_6 + SiH_4/He$		ArF		K, AES	[5.179]
Zn	$Zn(CH_3)_2/He$	SiO_2, Si	ArF, KrF	$W(\text{KrF}) \approx 10 \times W(\text{ArF})$	$\perp$, $\parallel$, EF	[5.184]
Zn	$Zn(CH_3)_2/He$	Glass, SiO_2, Si, Al	ArF, 257 Ar	$d \leq 0.7$	L, AL, K, SEM; nucleation, prenucleation technique	[5.8–10, 13, 185]
Zn	$Zn(CH_3)_2$	Glass	257 SH Ar		S, SEM, TEM; ripples	[5.12]
Zn	$Zn(CH_3)_2/Ar$	Al_2O_3, SiO_2	Hg/Xe lamp	$W = 1E-4$	EF, SEM, ESCA	[5.120]
Zn	$Zn(CH_3)_2/H_2$	(100) GaAs	ML Kr		S, L, T	[5.19, 20]
Zn	$Zn(C_2H_5)_2$	InP	ArF, KrF		L	[5.173]
Zn	$ZnSO_4$ + H_2SO_4/aq.	Si	HeNe, W lamp	$W = 0.01$ (0.01)	L; electroless plating	[5.66]
ZnO	$Zn(CH_3)_2 + NO_2$, $Zn(CH_3)_2$ + N_2O/He	Glass, SiO_2, Si	ArF, KrF	$W \leq 0.02$, $W = 3E-4$	$\perp$, $\parallel$, EF, HS 300–500, n, CP, $\rho = (0.1–1)E3$, AES, ESCA, X	[5.6, 28, 29, 184]
ZnO	ZnO (powder)	Glass, c-SiO_2, Al_2O_3, (100), (111) Si, (100), (111) GaAs, Au, Ti	CO_2		EF, HS 320–720, n, ρ, UV, VIS, IR, X; laser evaporation, (0002) oriented films	[5.186]
Zn_xSe_y	$Zn(CH_3)_2$ + $Se(CH_3)_2/Ar$	Al_2O_3, SiO_2	Hg/Xe lamp	$W = 1E-4$	EF, SEM	[5.120]

a Silver trifluoromethanesulfonate.
b Glass of complex or unspecified composition. Quartz glas., for example, is denoted by SiO_2, and crystalline quartz by c-SiO_2. Pyrex is a borosilicate glass. BK – 7 is delivered by Schott, Mainz, Germany.
c Refers to crystalline material; amorphous Si, for example, is denoted by a-Si.
d $Al(C_4H_9)_3$ is often abbreviated to TiBAl.
e AcAc is $[CH_3COCHCOCH_3]^-$ i.e. the 2,4-pentane dionato anion, which is also known as the acetylacetone anion.

f Sel-Rex (a division of Oxy-Metal Industr. Corp. Nutley, NJ 07110, USA). Autronex 55 GV, Au cyanide, no additives, 4 oz t Au/gal.
g Aurotron 439 N Schering, Berlin, Germany.
h Engelhard Industries, East Newark, NJ 07029, USA.
i HFAcAc is $[CF_3COCHCOCF_3]^-$ i. e. the 1,1,1,5,5,5 hexafluoro-2,4-pentane dionato anion, which is also known as the hexafluoroacetyl-acetonate anion.
j The adduct $(CH_3)_3InP(CH_3)_3$ dissociates at about 350 K.
k The exact composition of the bath can be found in [5.113].

5.1 Nucleation

The analysis of the microscopic mechanisms in the initial phase of growth is very complicated. In the phase of nucleation, we have to consider two cases: First, strongly absorbing substrates, and second, transparent substrates that do not absorb the laser light, or else do so only very slightly.

In the first case the molecules are thermally dissociated near the hot spot that is produced on the substrate by the absorbed laser light. The free atoms form clusters, which provide nucleation centers for further film growth. Under most processing conditions, the time of nucleation t_n is very small compared to the laser beam illumination times t_i that we shall henceforth be considering. The main differences from nucleation in large-area thin film growth techniques such as standard CVD ⌊5.187⌋, arise from the confinement of the temperature distribution and the related strong temperature gradients that are produced on the substrate surface. A further difference from CVD is based on the rapid change in the local laser-induced temperature distribution due to changes in reflectivity and thermal conductivity provided by the nuclei.

For transparent substrates, the situation is even more complex. Here, surface defects such as pinholes and scratches or dust particles may absorb the laser light and thereby allow nucleation to be initiated at certain places on the substrate surface. The mechanism is then very similar to that mentioned in the preceding paragraph. The situation becomes quite different when nucleation is initiated by atoms that result from nonthermal (single- or multiphoton) dissociation of adsorbed molecules (see Fig.5.1). It is clear

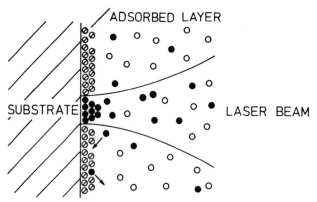

Fig.5.1. Schematic showing the influence of adsorbed-layer photochemistry on nucleation and condensation of gas-phase photofragments. Empty and crossed circles indicate free gas-phase and adsorbed precursor molecules, and full circles photofragments

that in such cases changes in the electronic properties of molecules due to adsorption, especially a shift and a broadening of the dissociative continuum, are of fundamental importance (Sect.3.1). Because of the high density of adsorbed molecules, the free atoms may form clusters, which, even when of subcritical size, may strongly absorb the laser radiation. Such heated clusters will then provide nucleation sites and film growth will proceed mainly thermally. In this case t_n is heavily dependent on the laser wavelength and may last several seconds or even minutes. The first evidence for such processes was obtained during deposition of Cd, Al and Ni from the alkyls and carbonyls, respectively ⌊5.10,14,65,105,106⌋. In any case, nucleation will cause not only a latent time, but also a threshold intensity for deposition.

In photolytic processing, where further film growth proceeds by mainly nonthermal dissociation of molecules, such multiatom clusters, irrespective of whether they were originally formed by thermal or nonthermal dissociation of species, may serve as nucleation centers for further film growth. An example is the deposition of metal microstructures produced by gas-phase dissociation of the corresponding alkyls or carbonyls by means of UV laser radiation (Sects.2.2.1 and 5.2). Here, in spite of the fact that metal atoms are produced within the total volume of the laser beam, condensation of atoms reaching the solid surface preferentially occurs at the nuclei produced within the area of the laser focus. This is schematically shown in Fig.5.1. Atoms that do not form stable nuclei (for Cd, the critical number of atoms, N_c, that is necessary for a stable nucleus is about 10 at 300 K; for nuclei adsorbed on substrate surfaces, this number also depends on the physical properties of the substrate) or attach themselves to these nucleation centers formed within the area of the laser focus, will diffuse across the surface and then evaporate. If an atom falls within a surface diffusion length $l = (2D\tau)^{1/2}$ (D is the surface diffusion coefficient of the atom and τ the characteristic time of reevaporation; the exact result depends on the size of the nucleus.) of a stable nucleus, it will, on average, be captured by this nucleus; if the atom falls outside this zone, it will reevaporate prior to capture. In other words, the sticking probability for free gas- or liquid-phase atoms or small clusters of atoms impinging on or near the nuclei formed within the irradiated substrate area is much larger than anywhere else on the substrate surface. Thus, nucleation thresholds improve the contrast in photolytic deposition, i.e. isotropic deposition occurs only to a small extent. The foregoing remarks show that we are a long way from an understanding of the initial dissociation and nucleation processes in laser-induced deposition.

5.2 Microstructures

In the following sections we will discuss pyrolytic and photolytic deposition of microstructures from adsorbed layers and from the gas, liquid and solid phases. For the reasons outlined in Sect.4.2, most of the experiments have been performed with cw lasers such as Ar^+ and Kr^+ lasers, including their harmonic lines.

5.2.1 Deposition from Adsorbed Layers

Some of the materials listed in Table 5.1 can be photochemically deposited from adsorbate-adsorbent systems under vacuum. Here, strong physisorption of the parent molecules, and little or no substrate heating by the incident laser light is required. Under these conditions, thermal desorption of the adsorbed molecules may be slow enough to allow photodeposition. This was investigated in some detail for Ni, Cd and Al deposited from metal-organic compounds on substrates such as glass, SiO_2 and Al_2O_3.

Nickel has been deposited by irradiating the substrates with 356 nm or 476 nm Kr^+ laser light $\lfloor 5.105,106 \rfloor$. The substrates were first exposed to a $Ni(CO)_4$ atmosphere of several millibars for 5-20 minutes. After this time several cycles of pumping and purging with He or H_2 followed. As expected from the increase in absorption of $Ni(CO)_4$, the latent times were much longer for 476 nm than for 356 nm radiation. Similarly, Cd and Al have been deposited from adsorbed $Cd(CH_3)_2$ and $Al_2(CH_3)_6$. These experiments were mainly performed with 193 nm ArF $\lfloor 5.4 \rfloor$, 257 nm frequency-doubled Ar^+ $\lfloor 5.14,60,61 \rfloor$, and 356 nm Kr^+ $\lfloor 5.65 \rfloor$ laser radiation. Very interesting information on the deposition of Al on $c\text{-}Al_2O_3$ has been obtained from pulsed optoacoustic (IR) spectroscopy $\lfloor 5.4 \rfloor$. These investigations have shown that high quantum yield nonthermal photodesorption of CH_3 groups incorporated in the already deposited Al films can be achieved with 193 nm ArF but not with 248 nm KrF laser radiation. Consequently, Al films grown from adsorbed layers with ArF laser light contain only low amounts of CH_3 contamination and are therefore of high quality. In any case, the thickness of layers deposited from adsorbate-adsorbent systems under vacuum is extremely small.

The contribution to the photodeposition rate of adsorbed layers that are in dynamic equilibrium with a surrounding gas phase has been studied for a number of systems $\lfloor 5.4,16,17,23,188 \rfloor$. As outlined in Sects.3.1 and 3.2, see (3.2) and (3.8,9), the relative importance of adsorbed-phase and gas-phase contributions can be discovered by investigating deposition rates as functions of laser focus diameter, substrate temperature and gas pressure.

The dependence of the photodeposition rate on the diameter of the laser focus has been studied for the deposition of Cd on SiO_2 substrates with 257 nm frequency-doubled Ar^+ laser radiation ⌊5.58,59⌋. From these investigations it was concluded that the decomposition of $Cd(CH_3)_2$ occurs mainly in the gas phase. For other systems, however, the importance of adsorbed-layer photochemistry for deposition rates was proved from the influence of the substrate temperature. An example is the deposition of Al from $Al_2(CH_3)_6$ on quartz substrates ⌊5.16,17,23⌋. Between 320 K and 280 K the deposition rate was found to increase with decreasing substrate temperature due to the increase in $Al_2(CH_3)_6$ coverage (Sect.3.1). Below the freezing point of the adsorbed layer, however, a decrease in the photodeposition rate was observed. This was interpreted in terms of an increase in photofragment recombination efficiency and by reduced molecular and atomic surface mobilities. Similar results have been obtained for the deposition of Ti from $TiCl_4$. In this latter system, however, the surface reaction is more complicated and seems to involve autocatalytic effects that arise from the deposited film itself. Further evidence for the importance of adsorbed layer photochemistry was obtained from the dependence of the Ti deposition rate on $TiCl_4$ pressure.

Multimolecular surface photochemistry was investigated for UV laser photo-polymerization of methyl methacrylate (MMA) into poly-MMA (PMMA) ⌊5.16,17⌋. This technique allows local deposition of moderate-molecular-weight PMMA. The photochemical mechanism is a free radical-catalyzed polymerization initiated by absorption of UV laser light by the volatile molecular layers that form on surfaces exposed to an ambient MMA vapor. The rapid collisions of vapor-phase molecules with the surface continually replenish the polymerizing adsorbed MMA and thereby permit rapid PMMA growth.

Photodissociation of adsorbed layers can also be used to prenucleate condensation areas, which are filled up later by large-area standard CVD or by pyrolytic LCVD. This combined technique allows maskless deposition of thin-film microstructures with much higher rates than those achieved in photolytic LCVD. Investigations of this kind have been performed mainly for Al deposited from $Al(C_4H_9)_3$ ⌊5.16,17⌋.

The influence of local electric field enhancement (Sect.3.1) due to surface contouring has been studied in particular for Cd deposition ⌊5.60,61⌋. These experiments used predeposited Cd spheres on C films. Cadmium was used for contouring because it possesses a plasma resonance in the appropriate ultraviolet spectral region. After covering the surface with a thin adlayer of $Cd(CH_3)_2$, which was in dynamic equilibrium with an atmosphere of 1.33 mbar $Cd(CH_3)_2$ and 1330 mbar Ar, the substrate was irradiated with 257 nm laser light. The growth pattern of the elliptical particles that grew from the spheres is in agreement with the theoretical model also presented in ⌊5.60,61⌋.

5.2.2 Gas-Phase Deposition

In this section we will discuss the production of microstructures in the form of spots, stripes and rods based on pyrolytic and photolytic laser-induced deposition from the gas phase, henceforth denoted by LCVD.

a) Pyrolytic LCVD

Investigations on the pyrolytic growth of spots are the simplest way of testing the adequacy of the model calculations presented in Sect.2.1.2 for the description of pyrolytic LCVD. The results also allow an understanding of the initial phase of the growth of rods (see below). For photolytic deposition, a quantitative or semiquantitative description has not yet been produced. Steady growth of stripes is relevant in pyrolytic and photolytic direct writing of microstructures. The following discussion concentrates on model systems for which the most complete data are available. Similar, but less detailed, investigations have been performed for the other materials listed in Table 5.1.

Deposition of Spots

When a Gaussian laser beam is focused onto an absorbing substrate (Fig.4.2a) that is immersed in one of the compounds listed in Table 5.1, and both the laser beam and the substrate are at rest, one initially observes the deposition of a circular spot. The most-detailed investigations on pyrolytic growth of spots have been performed by BÄUERLE and co-workers for Ni and Si deposited from $Ni(CO)_4$ and SiH_4, respectively.

Figures 5.2a,b show scanning electron micrographs of Ni spots deposited with 647 nm Kr^+ laser radiation. For the series of spots shown in Fig.5.2b, the laser power was held constant while the laser beam illumination time, t_i, was increased for each successive spot, working upwards from the bottom. The substrate material was glass covered with a 1000 Å a-Si layer. Such a substrate strongly absorbs the visible laser radiation and has a much smaller thermal conductivity than the deposit (Fig.2.11a). For such a substrate, the nucleation time t_n is very small compared to the laser beam illumination time t_i, i.e. $t_n \ll t_i$, and the spots grow very rapidly with t_i to radii much larger than the laser focus. Such large spot sizes can easily be measured and the high growth rates enable much data to be accumulated. Such experiments

Fig.5.2a,b. Scanning electron micrographs of Ni spots deposited from
Ni(CO)$_4$ by 647 nm Kr$^+$ laser radiation. The substrate was glass covered with
a 1000 Å a-Si layer. (a) Typical shape of a spot. (b) Series of spots
deposited at constant laser power with increasing laser beam illumination
time for each successive spot, working upwards from the bottom. The
micrograph was taken at glancing incidence (after [5.48,108])

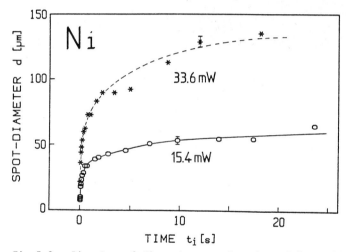

Fig.5.3. Diameter of Ni spots as a function of laser beam illumination time
for two powers of the 647 nm Kr$^+$ laser line. $2w_0 = 6$ μm, $p_{tot} \equiv p(Ni(CO)_4) =$
200 mbar. The broken curve is a guide for the eye. The full curve has been
calculated from (5.1) (after [5.48])

are therefore predestinated to be used to investigate the growth kinetics in
pyrolytic laser-induced deposition. This will now be outlined in greater
detail.

Figure 5.3 shows the diameter of Ni spots as a function of t_i for two
laser powers and a total gas pressure $p_{tot} \equiv p(Ni(CO)_4) = 200$ mbar
[5.48,108,190]. The figure shows that the condition $d > 2w_0$ (Sect.2.1.2) is

fulfilled for all but the shortest time in the 15.4 mW curve. The spot
diameter first increases very rapidly for a few tenths of a second and then
nearly saturates for times $t_i \approx 1$ to 10 s, depending on the laser power.
Because the temperature rise $\Delta T(d/2)$ decreases approximately inversely with
the spot diameter, see (2.19), this saturation can be explained by the
exponential dependence of the deposition rate on temperature, which yields an
apparent threshold below which deposition is negligible (Sect.3.2). However,
this saturation may also be due to a real threshold originating from
nucleation processes. The mechanism could be checked by depositing the same
material from reactant molecules with different activation energies. An
example is Si, which can be deposited either from SiH_4 or from $SiCl_4$.

Experimental results like those presented in Fig.5.3 enable the radial
growth rates $v \equiv \Delta d/2\Delta t$ of Ni spots to be calculated by numerical
differentiation. It is clear from the shape of the spots that the lateral
growth velocity v (oriented within the substrate surface) is not parallel to
the deposition rate W (defined as the rate of translation of a surface
element along its perpendicular) at the edge of the spot. However, because
spots grown with different laser powers remain similar in shape, W and v
differ only by a factor which is roughly constant. In any case, the
exponential dependence of the growth rate on temperature will dominate any
temperature dependence of the pre-exponential factor k_0 in (3.5). In other
words, if the growth of Ni spots is *thermally* activated, the lateral growth
velocity, v, should follow an Arrhenius type behavior. Figure 5.4 shows such
an Arrhenius plot in which additional experimental data for different laser
powers and laser focus diameters have been included. The plotted temperature
was that at the edge of the deposit, which was calculated as described in
Sect.2.1.2. The values used for the thermal conductivities correspond to
those in Fig.2.11a. The high temperature region of very fast growth
corresponds to small spot sizes, the low temperature region with much smaller
growth rates to larger spot sizes. The full line represents a least-squares
fit to the low temperature data and shows the exponential dependence of the
lateral growth rate on temperature. The apparent activation energy derived
from the figure is $\Delta E_1 = 22 \pm 3$ kcal/mole. This value is, within the accuracy
of the measurements, independent of P and w_0. This supports the idea that the
lateral growth rate depends only on the local temperature rise. The main
systematic error in the calculated temperatures, and consequently in the
activation energy, arises from the uncertainty regarding the reflectivity of
the deposited material, R_D, which was measured in situ and found to be $R_D = 0.2 \pm 0.1$. For the limiting values $R_D = 0.3$ and 0.1 we obtain $\Delta E_1 = 18$ and 31

88

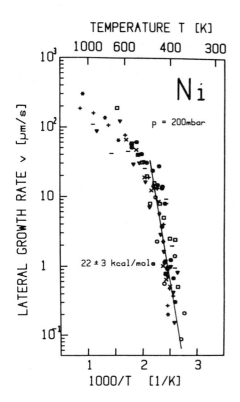

TEMPERATURE T [K]

Fig.5.4. Arrhenius plot for lateral growth rate v of Ni spots. The full line is a least-squares fit to the low temperature data. The different symbols refer to different laser powers and focus diameters (after ⌊5.108⌋)

kcal/mole, respectively. These values are within the range of apparent chemical activation energies reported for large-area heterogeneous deposition of Ni from $Ni(CO)_4$ ⌊5.191-194⌋. The investigations that covered the widest temperature region (350K ≤ T ≤ 430K) gave 22 kcal/mole ⌊5.192⌋. This latter value effectively corresponds to the energy required for the removal of the first CO group from the $Ni(CO)_4$ molecule and it is in excellent agreement with the activation energy derived from Fig.5.4. The marked decrease in slope appearing in the figure above the temperature $T_b \approx 500$ K could have various origins. For example it might indicate the limits of the model calculations, which are adequate only if $d > 2w_0$. Another possibility, which at present seems very likely, is the change in reflectance of the Ni spot that may occur for the very shortest times t_i. However, the decrease in slope could also indicate limitations in the gas-phase transport of reactant molecules or reaction products or in the desorption of CO (Sects.3.1,2). Another reason could be that other chemical reaction pathways, characterized by an apparent activation energy ΔE_2, may become possible at higher temperatures.

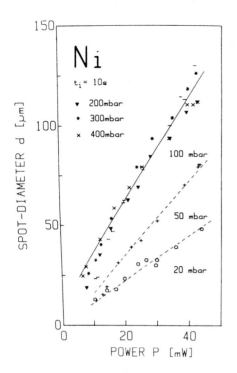

Fig.5.5. Dependence of Ni spot diameter on laser power for various gas pressures and constant laser beam illumination time t_i = 10 s. λ = 647 nm Kr$^+$, $2w_0$ = 6 μm. The <u>full line</u> was calculated from (5.1). The <u>dashed lines</u> are guides for the eye (after [5.108])

Figure 5.5 shows the diameter of Ni spots for constant laser beam illumination time t_i = 10 s, as a function of laser power and for different gas pressures. The saturation in growth observed for higher gas pressures is expected for a heterogeneous chemical reaction and is described by the pressure dependence of k_0 in (3.5). From Fig.5.5 we find $k_0 \approx 1$ for $p \geqslant 100$ mbar. The diameter of Ni spots can be calculated in a comparable way to (3.6) from

$$d(t_i) = d(t_m) + 2A \int_{t_m}^{t_i} dt\ k_0(p,T(t))\ \exp\lfloor-\Delta E_j/RT(t)\rfloor\ . \tag{5.1}$$

Here, $\Delta E_j \equiv \Delta E_1$ and $\Delta E_j \equiv \Delta E_2$ are the apparent activation energies in the low and high temperature regions, i.e. below and above T_b, respectively. It is clear from Sect.2.1.2 that the temperature T itself depends on the spot diameter and is therefore a function of time. The lower integration limit in (5.1) is the time for which the assumption $d(t_m) > 2w_0$ is fulfilled (Sect.2.1.2). For all but the lowest laser powers we find $t_m \ll t_i$ = 10 s. For p = 200 mbar, we can derive the activation energies and the constant A from Fig.5.4. Integration of (5.1) can be performed by assuming $d(t_m) \approx 2w_0$. Note that $d(t_i) \gg 2w_0$. The integration shows that d increases quasi-linearly with laser power for P $\leqslant$ 60 mW. This result is included in Fig.5.5 by the full curve. The agreement with the experimental data is excellent and shows the consistency of the procedure. It can easily be shown that contrary to the value of ΔE_1, the (larger) uncertainty in ΔE_2 does not appreciably influence the result. A further proof of the consistency of the procedure is obtained when calculating from (5.1) the time dependence of the spot diameter (full curve in Fig.5.3).

We now compare the pressure dependence of the spot diameter for constant t_i and constant laser power (Fig.5.5) with the reaction kinetics derived by CARLTON and OXLEY [5.192]. Here, it is essential to take into account that in the experiments described above, the partial pressure of CO is negligible compared to that of $Ni(CO)_4$. According to [5.192] the pre-exponential factor in the rate equation (3.5) is then given by

$$k_0(p(Ni(CO)_4),T) = k^2 p^2/(1 + kp)^2 , \qquad (5.2)$$

with $k = 0.067 \exp(T_1/T)$ mbar^{-1} and $T_1 = 352$ K. Saturation should occur if $kp \gg 1$. For a laser power of $P = 30$ mW, saturation of the spot diameter would therefore be expected (within 3%) for a pressure of $p = 50$ mbar. Clearly, only a qualitative agreement with the experimentally observed pressure dependence shown in Fig.5.5 can be expected, for several reasons: the difference between W and v enters the pre-exponential factor that has been determined only for $p = 200$ mbar (Fig.5.4); the factor k depends on temperature and thereby on the diameter of the spot. The apparent and perhaps even the true order of reaction may change with pressure - as well as with temperature. This may also explain the considerable differences reported in the literature for the form of k_0 and the entering parameters (see (5.2) and [5.191-194]).

Steady Growth of Rods

As outlined above, radial growth of pyrolytically deposited spots saturates with laser beam illumination time t_i (Fig.5.3). Simultaneously with this saturation, an increase in axial growth is observed. Consequently, for longer times t_i, the growth of a rod along the axis of the laser focus is observed. The growth of rods has been investigated for various different materials by BÄUERLE et al. [5.48,49,51,104,105,123,124,126,170,195]. A typical example of a rod is shown in Fig.5.6 for the case of Si deposited from SiH_4. Two phases of growth can be observed. Near the onset, the deposition rate depends strongly on the physical properties of the substrate. This phase of growth corresponds with the growth of spots and has already been discussed. In steady growth, which is characterized in Fig.5.6 by a constant rod diameter,

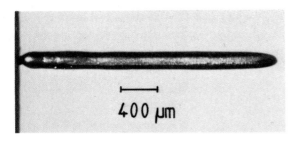

Fig.5.6. Silicon rod grown from SiH_4 with 488 nm Ar^+ laser radiation. P = 400 mW, $p_{tot} \equiv p(SiH_4)$ = 133 mbar (after ⌊5.50⌋)

the deposition rate is independent of the substrate material. Therefore, in contrast to the growth of spots, the temperature profile in the tip of the rod is independent of time, i.e. T = T(r), if the laser irradiance is held constant. The constant rod diameter is a consequence of the threshold in lateral growth - as discussed above. It is clear from the shape of the rod that in steady growth, the maximum growth rate is at the center of the tip and identical with the axial growth velocity. Therefore, the deposition rate can be defined by the growth of the length of the rod per unit time

$$W(T) \equiv v(r = 0,T) = \Delta h(r = 0,T)/\Delta t \;. \tag{5.3}$$

The surface temperature at the center of the tip of the rod, T, can be measured as described in Sect.4.4 (see also ⌊5.51,196⌋). Figure 5.7 shows an Arrhenius plot for the deposition of Si from SiH_4. The upper curve refers to data obtained from pyrolytic LCVD of rods. In the kinetically controlled regime, which reaches up to about 1400 K, the deposition rate increases exponentially with temperature and is characterized by an apparent chemical activation energy of ΔE = 44 ± 4 kcal/mole ⌊this value is *not* corrected for the temperature dependence of k_0 (see (3.5) and ⌊5.51⌋)⌋. The characteristic decrease in slope observed above a certain temperature may indicate that depbsition is no longer controlled by the chemical kinetics, but instead becomes limited by transport (Sect.3.2). However, alternative explanations cannot be ruled out (see discussion of Fig.5.4). The lower part of Fig.5.7 shows the deposition rate for Si deposited from SiH_4 - with H_2 as carrier gas ⌊$p(SiH_4)$ = 1 mbar, p_{tot} = 1000 mbar⌋ - according to standard CVD techniques ⌊5.197⌋. The comparison of LCVD and CVD curves shows the remarkable differences between small- and large-area chemical reactions, as discussed in Sect.3.2.

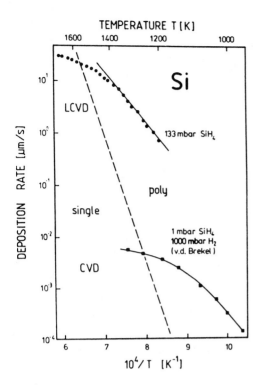

TEMPERATURE T [K]

DEPOSITION RATE [μm/s]

10⁴/T [K⁻¹]

Si

LCVD

133 mbar SiH₄

poly

single

1 mbar SiH₄
1000 mbar H₂
(v.d. Brekel)

CVD

Fig.5.7. Arrhenius plot for deposition of Si from SiH₄ in LCVD and CVD. The broken line separates regions of single- and polycrystalline growth (after ⌊5.104⌋)

It proved possible to vary the diameter of Si rods between 20 μm and 380 μm for effective laser powers of 25 mW and 1.4 W, respectively, of the λ = 488 nm Ar⁺ laser line. The lower limit was essentially determined by the mechanical stability of the apparatus, the upper limit by the available laser power. Due to the step-like increase in reflectivity at the melting point, the total absorbed laser power, $P(1-R_D)$, falls dramatically at this temperature.

Similar investigations were performed for the deposition of C from C_2H_2, C_2H_4, C_2H_6 and CH_4 ⌊5.49,51-53⌋. The apparent chemical activation energy derived for the deposition from C_2H_2 was ΔE = 47.3 ± 0.6 kcal/mole (uncorrected value). This activation energy was independent of the gas pressure and temperature within the investigated ranges of 50 mbar ⩽ $p(C_2H_2)$ ⩽ 1000 mbar and 1900 K ⩽ T ⩽ 2450 K. The reaction order was found to be γ = 0.8. The corresponding values for C_2H_4 were ΔE = 58.3 ± 1.3 kcal/mole for 300 mbar ⩽ $p(C_2H_4)$ ⩽ 1000 mbar, 2000 K ⩽ T ⩽ 2250 K and γ = 0.8. For C_2H_6, ΔE = 78.9 ± 4 kcal/mole, 300 mbar ⩽ $p(C_2H_6)$ ⩽ 1000 mbar, 2200 K ⩽ T ⩽ 2650 K, γ = 2. For CH_4, ΔE = 119 ± 2 kcal/mole for 500 mbar ⩽ $p(CH_4)$ ⩽ 1000 mbar, 2850 K ⩽ T ⩽ 3100 K, γ = 1.25, and ΔE = 43.5 ± 1.4 kcal/mole for 500 mbar ⩽ $p(CH_4)$ ⩽ 1000 mbar, 2400 K ⩽ T ⩽ 2750 K, γ = 2.2.

93

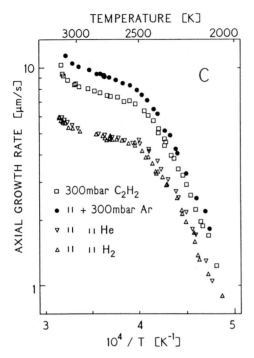

TEMPERATURE [K]

AXIAL GROWTH RATE [μm/s]

10^4 / T [K^{-1}]

□ 300mbar C$_2$H$_2$

● ΙΙ + 300mbar Ar

▽ ΙΙ   ΙΙ He

△ ΙΙ   ΙΙ H$_2$

C

Fig.5.8. Arrhenius plot for the laser-induced deposition of carbon from pure C$_2$H$_2$ and from gas mixtures of C$_2$H$_2$ with H$_2$, He and Ar (after [5.53])

   The influence of thermal diffusion on the deposition rate of C was investigated in detail for gas mixtures of C$_2$H$_2$ with H$_2$, He and Ar [5.53]. Figure 5.8 shows the results. Due to thermal diffusion, the lighter gases (He and H$_2$) accumulate near the hot tip of the rod, thus reducing the partial pressure of the reactant and thereby the deposition rate. Ar, on the other hand, has a higher molecular weight than C$_2$H$_2$. It is therefore depleted near the hot surface and, consequently, increases the partial pressure of C$_2$H$_2$, and thereby the deposition rate. The figure clearly shows that the selection of carrier gases used in LCVD has an important influence on the deposition rates achieved.

   Rods of SiO$_x$ and of stoichiometric SiO$_2$ were grown by using λ = 530.9 nm Kr$^+$ laser radiation and a mixture of N$_2$O and SiH$_4$ [5.170]. The growth of Ni rods from gaseous Ni(CO)$_4$ was investigated for laser powers ranging from about 10 mW up to about 200 mW [5.105]. The deposition rate did not change when Kr$^+$ laser wavelengths between λ = 476.2 nm and 647.1 nm were used (Table 4.1) - as one would expect for a thermally activated process (see Sect.2.1 and Chap.3). The limit of growth up to which controlled deposition was possible, was set by the occurrence of spontaneous breakdowns, probably due to an autocatalyzed reaction, within the gas phase above the surface of the tip of the rod.

Investigations of the temperature dependence of the deposition rate during the steady growth of rods are the most accurate method so far available for obtaining information on the reaction kinetics in pyrolytic LCVD. Furthermore, because of the extremely high deposition rates together with the possibility of in situ temperature measurements, this technique seems to be unique for rapid and accurate determination of apparent chemical activation energies that are also relevant to CVD and gas-phase epitaxial processes. The determination of such activation energies by the standard techniques is very time-consuming and problematic because a number of parameters, such as substrate temperature, gas velocity and gas mixture, must be held constant over long periods of time, generally several hours, and only small numbers of data points can be generated ⌊5.51⌋.

Other interesting possibilities, such as the growth of complicated three-dimensional structures by interference of laser beams, the production of materials with higher purity and the production of nonequilibrium materials or materials which form only under extreme conditions, can only be speculated upon.

The microstructure of rods has been mainly investigated by optical microscopy, scanning electron microscopy (SEM), X-ray diffraction and Raman scattering. The microstructure depends on the laser-induced temperature and the gas pressure. Rods have been grown in amorphous (SiO_x, SiO_2), polycrystalline (Ni, C, Si) and single-crystalline phases (Si). In this paragraph, we will only discuss further the growth of single-crystal Si rods ⌊5.48,104,124⌋. Figure 5.9 shows a scanning electron micrograph of the tip of

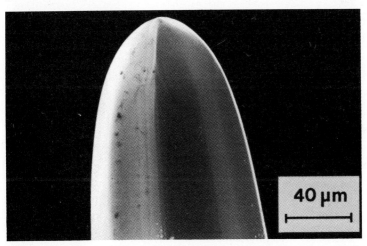

Fig.5.9. Scanning electron micrograph of the tip of a single-crystal Si rod grown at 1650 K with 530.9 nm Kr^+ laser radiation. $p(SiH_4)$ = 133 mbar (after ⌊5.124⌋)

such a rod that was grown from SiH_4 at 1650 K with 530.9 nm Kr^+ laser radiation. The orientation of the axis of such rods was found to be close to either <100> or <110> directions and independent of the substrate. Note that these are the fastest directions of growth in crystalline Si. For the silane pressure used, $p_{tot} \equiv p(SiH_4)$ = 133 mbar, single-crystal growth was observed only above 1550 to 1650 K. In this connection it is interesting to recall the microstructure of Si films grown on single-crystal Si substrates by standard CVD techniques. Here, it has been found that the regime of polycrystalline growth is separated from the regime of single-crystal growth by a border line (dashed line in Fig.5.7) which is essentially determined by the ratio of the flux of Si atoms giving rise to the observed growth rate, and the value of the self-diffusion coefficient of Si needed to arrange the arriving atoms on proper lattice sites. Linear extrapolation of this border line to higher temperatures yields an intersection point with the LCVD curve at about 1520 K. This value is in remarkable agreement with the temperature limit we find for single-crystal growth of rods.

Pyrolytic Direct Writing

Pyrolytic direct writing was investigated by translating the substrate in one dimension perpendicular to the focused laser beam. In this case one obtains stripes. The most-detailed investigations have been performed for Ni and Si.

As in the case of spots and rods, the morphology of stripes strongly varies with laser power ⌊5.48,50,104,105,198⌋. This is shown in Figs.5.10a-d for Si stripes deposited from SiH_4 on Si wafers. Here, 488 nm Ar^+ laser radiation was used. At laser powers corresponding to center temperatures below the melting point of Si, T_m = 1683 K, a convex cross section is observed, and the situation is very similar to that discussed below for direct writing of Ni stripes. When the laser power is further increased, a dip in the middle of the stripe (Fig.5.10c) occurs for center temperatures $T > T_m$. This dip further increases with increasing laser power (Fig.5.10d). The dip can be explained by the change in surface tension with temperature, which pulls the liquid away from the valleys. At such laser powers, polysilanes, probably formed in a homogeneous reaction above the surface of the deposit, condense in the region of deposition. Melting of the surface of the stripe is accompanied by the occurrence of a ripple structure with grating vector **k** parallel to the incident electric field vector **F**. The formation of ripples and of periodic superstructures that occur under certain experimental conditions will be discussed in Sect.5.2.5.

The influence of melting and changes in the morphology of stripes complicate the understanding of the deposition process. The following

96

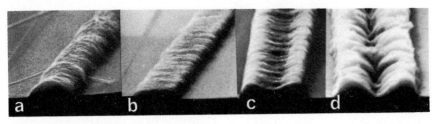

Fig.5.10a-d. Stripes of Si grown on Si wafers with increasing laser power (left to right). $p(SiH_4) \approx$ 40 mbar, $v_s \approx$ 10 µm/s, λ = 488 nm (after [5.198])

 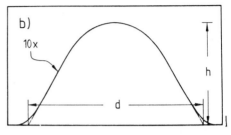

Fig.5.11a,b. (a) Ni stripe grown on glass covered with 1000 Å a-Si. P = 6 mW, $2w_0$ = 2.5 µm, λ = 530.9 nm, $p(Ni(CO)_4)$ = 400 mbar, scanning velocity v_s = 84 µm/s. (b) Typical thickness profile of a Ni stripe. The vertical scale is expanded ten times (after [5.105])

analysis of direct writing will therefore be confined to low laser powers where no appreciable changes in the shape of the cross section occur and where, therefore, an unequivocal definition of a stripe width and height is possible. The main experimental results will be outlined in detail for the example of Ni stripes that have been deposited on substrates of different absorbance and thermal conductivity [5.105,106].

A typical Ni stripe together with its interferometrically measured thickness profile, is shown in Fig.5.11. We shall henceforth define the width of the stripes by d, and the thickness by h.

Figure 5.12 shows the width and thickness of Ni stripes as a function of laser power for three different substrate materials. In all cases, the cross section of the stripes is similar to that shown in Fig.5.11. The height, measured in the middle of the stripe, is typically h ⩽ (0.05-0.1)d. The negligible influence of the laser wavelength (Fig.5.12b) supports the idea that the decomposition mechanism is mainly thermal. Note that the results for the 1000 Å a-Si/glass substrate and the uncovered glass substrate are approximately equal (Figs.5.12a,b). Both substrates have about the same thermal conductivity. However, their absorbances for 530.9 nm Kr^+ laser radiation differ by several orders of magnitude. Therefore, these results strongly suggest that in steady growth of stripes the total laser power

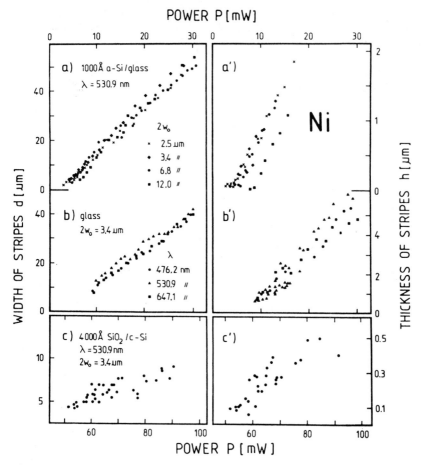

Fig.5.12. Dependence of width and thickness of Ni stripes on total laser power for different focus diameters, wavelengths and substrates. In all cases, the total pressure was $p_{tot} \equiv p(Ni(CO)_4) = 400$ mbar, and the scanning velocity = 84 μm/s (after [5.105])

absorbed is determined by the absorbance of the already deposited material and is given by $P(1-R_D)$. Apparently, steady growth of stripes occurs as drawn schematically in Fig.5.13. During the dwell time of the laser beam, lateral growth occurs. In steady growth, the lateral growth velocity in the scanning direction, v, must be equal to the scanning velocity of the laser beam, v_s, i.e. $v = v_s$ (this condition does not hold in explosive deposition, which is briefly discussed in Sect.5.2.5). Steady growth of stripes tails off when the scanning velocity of the laser beam exceeds the maximum lateral growth velocity, i.e. if $v_s > v_{max}$. In fact, the maximum lateral growth rates found for the deposition of Ni spots on 1000 Å a-Si/glass agree, to within a factor

98

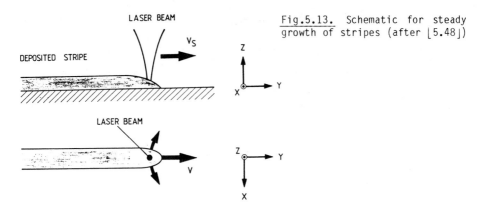

Fig.5.13. Schematic for steady growth of stripes (after ⌊5.48⌋)

of about two, with the maximum scanning velocities obtained for the same system and the same set of parameters (see below).

Comparison with Model Calculations

We now compare the results of Fig.5.12 with model calculations presented in Sect.2.1.2. Assume we produce the same center temperature T_c for all three substrate materials considered. From the full curves in Figs.2.12a-c it then becomes clear that in cases b and c the temperature falls off much faster away from the center than in case a (note that the temperature distributions shown in Fig.2.12 have been calculated for circular cylinders; however, as outlined in Sect.2.1.2, these temperature profiles are very similar to those obtained within cross sections of stripes). Because growth occurs only down to a threshold temperature T_t (see also Sect.5.1), the stripe in Fig.2.12b or c grows to a final width much smaller than that in Fig.2.12a. This explains the different width of stripes in Figs.5.12a-c. For thermally insulating substrates, the width of metallic stripes, here the Ni stripes (Figs.5.12a,b), grow to a final width that is much larger than the laser focus, i.e. $d \gg 2w_0$ (except for the lowest laser powers). On the other hand, if the thermal conductivities of the deposit and the substrate are equal, as in the case of Fig.5.12c, the width of stripes remains of the order of the diameter of the laser focus. The results of the model calculations presented in Fig.2.12a also explain another feature in Fig.5.12a. For medium and higher laser powers (for these powers under consideration $T_c < T_m(Ni)$, and no changes in the shape of the cross section have been observed) the widths of stripes are independent of the diameter of the laser focus. Let us consider, for example, the effect on the temperature distribution (dash-dotted curves in Fig.2.12) of doubling the laser focus diameter at constant laser power. In the case of Fig.2.12a, the temperature rise $\Delta T(d/2)$ remains approximately the

same. As a consequence, the width of the stripe is expected to remain
unaffected, in agreement with the results of Fig.5.12a. However, in
Figs.2.12b,c, deposition becomes impossible when the laser focus is doubled,
because even the center temperature T_c decreases below the threshold
temperature for deposition. If in the case of Fig.2.12a T_c approaches T_t,
deposition will continue to lower laser powers as the diameter of the laser
focus becomes smaller. This is reflected also in the experimental results of
Fig.5.12a. In other words, the smallest widths of structures that can be
achieved, depend on $2w_0$. The fact that stripes can be produced that are
narrower than the diffraction limit of the optical system (see Fig.5.12a)
originates from the threshold for deposition, discussed above (see also
Sects.3.5 and 5.1).

The dependence of the width and thickness of stripes on the scanning
velocity is shown in Fig.5.14. For the parameters used in these experiments,
the breakoff occurs at about 130 μm/s. When increasing P, under otherwise
identical experimental conditions, the scanning velocity can also be
increased. The dashed curve in Fig.5.14 was calculated by using v =
$v_0 \exp(-\Delta E/RT)$ and $T = T_0 + \Delta T$, together with (2.19) and the steady state

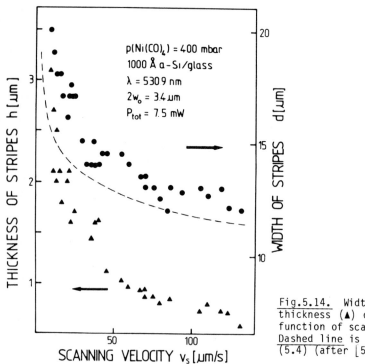

Fig.5.14. Width (•) and thickness (▲) of stripes as a function of scanning velocity. Dashed line is calculated from (5.4) (after [5.105])

condition $v = v_s$, giving

$$d = \frac{P\,(1 - R_D)}{2\kappa\,\{(\Delta E/R)\lfloor \ln(v_0/v_s) - T_0\rfloor\}} \,. \tag{5.4}$$

Here, $v_0 = Ak$ has been taken from the Arrhenius plot in Fig.5.4. In other words, the dashed curve in Fig.5.14 contains no additional parameter. Comparison with the experimental data shows that this crude approximation describes the absolute width as well as its dependence on the scanning velocity reasonably well. The explanation of the remaining deviations requires a self-consistent treatment, the consideration of the velocity dependence of the exact position of the laser focus with respect to the tip of the growing stripe and of the heat transport along the stripe (Sect.2.1.2). A more sophisticated model, which takes these effects into account, makes it possible to describe the experimental data quantitatively ⌊5.190⌋.

The foregoing results are by no means specific to direct writing of Ni stripes but are characteristic of pyrolytic LCVD of flat structures in general. Similar results corresponding to the case $\kappa_D \gg \kappa_S$, were obtained for the deposition of Al and Si on 1000 Å a-Si/glass ⌊5.125,198⌋, for C on Al_2O_3 ⌊5.49⌋, and with some restrictions, for Cu on Si ⌊5.74⌋. In addition to the example of Ni on 4000 Å SiO_2/Si (Fig.5.12c), the case $\kappa_D \approx \kappa_S \gg \kappa_L$ (Fig.2.12b) applies also to Si on 4000 Å SiO_2/Si ⌊5.128,129⌋. The case $\kappa_D = \kappa_S = \kappa_L$ has been verified for Ni on Si ⌊5.106,107⌋ and for Si on Si ⌊5.126,128,129⌋. In these latter cases the widths of stripes are typically $d \lesssim 2w_0$. This is what Figs.2.12b,c would lead us to expect.

From the above discussion it is clear that many features in direct writing can be qualitatively or semiquantitatively understood from calculated temperature profiles as presented in Sect.2.1.2. The foregoing results also show that in direct writing the range of parameters and therefore the lateral and axial growth rates and the related maximal scanning velocities strongly depend on both the physical properties of the deposited material *and* of the substrate. While the possible range of variation in the width of stripes is very large for $\kappa_D \gg \kappa_S$, it is very small for $\kappa_D \approx \kappa_S$. The upper limit is essentially based on the maximum center temperature at which controlled deposition is possible, i.e. no dramatic changes occur in the geometry of the deposit, there is no damaging of the substrate and no triggering of a homogeneous gas-phase reaction above the surface of the deposit. Furthermore, small changes in $2w_0$ or in the positioning of the substrate or in the laser power (due to systematic uncertainties or due to mechanical or electrical instabilities) will have a much stronger influence for systems where $\kappa_D \approx \kappa_S$ than for those where $\kappa_D \gg \kappa_S$. This may also explain the larger scatter in the data of Fig.5.12c compared to those in Figs.5.12a,b.

Pyrolytic direct writing of electrically highly conducting B doped Si stripes with linewidths as small as 1 μm was first reported by EHRLICH et al. ⌊5.129⌋. The dopant gases were BCl_3 and $B(CH_3)_3$. Similar experiments using $Al_2(CH_3)_6$ as dopant were unsatisfactory. The reason is probably the large difference in pyrolytic decomposition rates of $Al_2(CH_3)_6$ and SiH_4. With $B(CH_3)_3$ the best films had resistivities of about $1.5 \cdot 10^{-3}$ Ωcm.

Deposition with IR Lasers
Pyrolytic deposition of structures in the form of spots, rods, and stripes by means of IR light was mainly performed with 10.6 μm cw or pulsed CO_2 laser radiation (Table 5.1). First experiments have been reported for C ⌊5.55⌋ and for Si ⌊5.136⌋. Detailed investigations on Si deposition have been performed by HANABUSA et al. ⌊5.149,150⌋. Because the diffraction-limited diameter of the laser focus is proportional to the wavelength of the light λ, see (4.2), the smallest lateral dimensions of deposits are much larger, typically by a factor of 20-50, than those achieved with visible or ultraviolet light (Sect.3.5). Consequently, mass transport will often be determined by one-dimensional diffusion. This significantly reduces the maximum (thickness) deposition rates with respect to those achieved with visible or ultaviolet light. The formation of thin extended films by means of IR laser radiation will be outlined in Sect.5.3.

b) Photolytic LCVD

Production of microstructures by photolytic LCVD based on the excitation of dissociative electronic transitions of parent molecules (Sect.2.2.1) has been studied mainly for metals. The precursors most commonly used are metal-alkyls, metal-carbonyls and metal-fluorides (Table 5.1). The microscopic mechanisms for decomposition are based on single-photon or multiphoton processes.

Single-photon decomposition has been most thoroughly studied for $Cd(CH_3)_2$, and $Al_2(CH_3)_6$ ⌊5.8⌋. These compounds were preferred because they are readily available, have a relatively high vapor pressure and, very importantly, they show a dissociative continuum in the near to medium UV that can be easily reached by frequency doubling of cw Ar^+ or Kr^+ laser lines, by excimer lasers, or by harmonic lines of Nd:YAG lasers. Furthermore, for $Cd(CH_3)_2$ some of the reaction channels were already known for the homogeneous photodecomposition process (Sect.2.2.1). Therefore, $Cd(CH_3)_2$ can be considered as a model substance for investigating photolytic deposition based on single-photon decomposition processes. Dissociation of the molecule proceeds according to (2.20), and, of course, takes place within the total volume of the laser beam. This reaction, or ones similar to it, may therefore be used to produce extended thin films at low substrate temperatures by using either perpendicular or parallel laser irradiation (see Sect.5.3 and Figs.4.2b,3). Nevertheless, it has been shown that photolytic processing also allows single-step production of microstructures. The spatial confinement of the deposition process is probably closely related to physisorbed layers of the parent molecules and to the nucleation process itself. This has already been outlined in Sects.3.1, 3.5 and 5.1. In cases where physisorption is very weak, severe problems with the confinement of the reaction may occur [5.72]. On the other hand, it is possible to achieve good localization and feature contrast even in cases such as these by taking a number of experimental precautions - as already discussed in Sect.3.5.

Figure 5.15 shows the (thickness) deposition rate for Cd spots as a function of the intensity of the 257.2 nm frequency-doubled Ar^+ laser line. The linear increase in deposition rate with laser power is expected for single-photon dissociation (Sects.2.2.1 and 3.2). At low laser fluences the threshold for surface nucleation prevents deposition. At much higher laser fluences than those shown in Fig.5.15, the deposition rate saturates due to mass transport limitations. This effect is more pronounced when a buffer gas

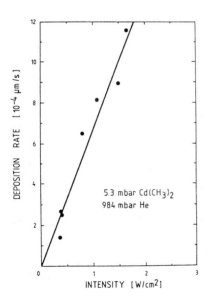

5.3 mbar $Cd(CH_3)_2$
984 mbar He

Fig.5.15. Deposition rate for Cd spots versus 257 nm second harmonic Ar^+ laser light intensity. $p(Cd(CH_3)_2)$ = 5.3 mbar. He was used as a buffer (after [5.8])

is used. Mass transport could be increased by forced gas convection or a flow. Deposition rates for Cd of up to 0.1 μm/s have been measured at UV intensities of about 10^4 W/cm^2.

The absolute value of the deposition rate depends on the absorption cross section σ_a of the molecule at the laser wavelength, on the partial gas pressure of the reactant and, when a buffer gas is used, on the total gas pressure. For example, at 257.2 nm the absorption cross section for $Al_2(CH_3)_6$ is smaller than that for $Cd(CH_3)_2$ by more than a factor of 10^3, resulting in a corresponding decrease in deposition rate at otherwise constant conditions (Table 2.1). This example shows one of the main limitations for photolytic LCVD based on single-photon processes. Deposition rates that are attractive for applications of the technique can only be reached when the laser wavelength matches a dissociative transition of the molecule. For many molecules which would be suitable for the deposition of metals, semiconductors, or insulators, these transitions are located in the medium to far ultraviolet, where, at present, only a small number of comparatively intense laser light sources are available (Table 4.1). A more fundamental, and therefore even more severe, limitation is the tendency for homogeneous cluster formation at higher laser irradiances and/or partial pressures of reactants. Clusters may condense everywhere on the substrate surface, the entrance windows, etc. Controlled deposition then becomes impossible. For these reasons, the deposition rates in photolytic LCVD will be always much lower than those achieved in pyrolytic LCVD.

Controlled growth in UV laser photodeposition of metals was observed for laser irradiances of, typically, 1 to 10^4 W/cm^2, and for gas pressures extending from 0.1 to 100 mbar. Typical deposition rates achieved were 10^{-3} to some 0.1 μm/s (Table 5.1). These deposition rates are a factor of 10^4-10^2 smaller than those obtained in pyrolytic LCVD. The main advantages of photolytic LCVD are the lower local processing temperatures and the smaller influence of the physical properties of the substrate. On the other hand, without uniform substrate heating, the microstructure, purity, and electrical properties of photodeposited films are, in many cases, unsatisfactory (Sect.5.4). The lateral resolution achieved in photolytic deposition of stripes was of the order of some 0.1 μm (Sect.3.5).

Coherent multiphoton decomposition processes based on dissociative electronic excitations are only of limited value for controlled deposition of well-defined micrometer-sized structures. The main reasons have already been outlined in Sect.4.2. In fact, detailed investigations into the deposition of Cr from $Cr(CO)_6$ by MAYER et al. ⌊5.72⌋ have illuminated many of these problems. Additionally, the Cr films had poor adherence and contained a large amount of C. The adherence of films could probably be improved by heating the substrate either during the deposition process or prior to removal from the reaction chamber.

There is as yet *no* clear example of photolytic LCVD based on selective multiphoton *vibrational* excitations with IR laser radiation. The only detailed investigations have concentrated on the deposition of Si from SiH_4 by means of pulsed CO_2 laser radiation $\lfloor 5.137,149,150,153 \rfloor$. However, the decomposition mechanisms seem to be dominated by gas-phase heating and not by multiphoton processes (see also Chap.2 and Sect.5.3).

c) Combined Pyrolytic-Photolytic (Hybrid) LCVD

Pyrolytic and photolytic LCVD have both their characteristic advantages and disadvantages: The deposition rates in pyrolytic LCVD exceed those in photolytic LCVD by several orders of magnitude. Consequently, pyrolytic LCVD allows much higher scanning velocities in direct writing and also the production of three-dimensional structures. The microstructure and the electrical properties of pyrolytically deposited materials are in general superior to those deposited by photolysis (Sect.5.4). Laser pyrolysis is much more universal in the sense that the reaction rate depends only slightly on the exact wavelength of the laser light. For this reason, a great variety of materials can be deposited. The basic advantages of photolytic LCVD are the lower local processing temperatures and the lower sensitivity to the physical properties of the substrate. Therefore, direct writing of patterns over different materials with significantly changing thermal properties can be performed more uniformly.

A possibility that makes the best of the advantages and disadvantages of both pyrolytic and photolytic LCVD is a twin-beam or single-beam combined pyrolytic-photolytic reaction. Initial investigations of this type have been performed for the deposition of Ni from $Ni(CO)_4$ $\lfloor 5.50,104-106 \rfloor$. Here, it was clearly demonstrated that for (visible) laser radiation that is absorbed neither within gaseous $Ni(CO)_4$ nor by the substrate, the latent times for nucleation (Sect.5.1) are significantly reduced when the UV plasma radiation of the laser $\lfloor$ this is absorbed by $Ni(CO)_4 \rfloor$ is not blocked but focused onto the substrate together with the laser light. Similar results have been obtained when using 356 nm Kr^+ laser radiation only. This wavelength is slightly absorbed by $Ni(CO)_4$, but not by quartz substrates. In both cases, when nucleation was started, deposition proceeded mainly pyrolytically and

was then essentially equal for the different Kr^+ laser lines between 356 and
647 nm (Table 4.1), and independent of the presence of the plasma radiation.
This can be understood from the strong absorption of the deposited Ni, which
is approximately constant within this range of wavelengths. Hybrid deposition
was recently investigated in greater detail for Mo, W and Pt by GILGEN et al.
[5.103]. Here, the UV multiline Ar^+ laser output between 351 and 364 nm was
used together with $MO(CO)_6$, $W(CO)_6$ and $Pt(HFAcAc)$ as parent molecules. These
experiments have confirmed that metal films of good morphology and electrical
properties (Sect.5.4) can be deposited without uniform substrate heating or
curing only by laser pyrolysis. Nevertheless, photolytic mechanisms that also
contribute at these laser wavelengths allow direct patterning of transparent
substrates such as glass or sapphire, without a measurable initial delay or
latent time. Clearly, photolysis is relevant here only for initiation of the
deposition process, because afterwards absorption is determined by the
already deposited material (see above and Sect.2.1.2).

5.2.3 Liquid-Phase Deposition, Electroplating

Laser-induced deposition from the liquid phase has been demonstrated mainly
for metals. KARLICEK et al. [5.42] have deposited Pt, Au, and Ni from aqueous
H_2PtCl_6, $HAuCl_4$ and $NiSO_4$ solutions, respectively. Also, Pt and Au were
deposited from the corresponding methanol solutions. The substrate materials
used were mainly p-type, n-type and undoped (110) and (111) InP and, for Pt
deposition, also n-type GaAs. Pulsed dye laser radiation ranging from 580 nm
to 720 nm was employed (within this wavelength range, the metal salt
solutions are transparent). The light intensities were sufficient to initiate
thermal decomposition of the InP surface, but below the threshold at which
gross surface damage occurs. Smooth platinum films up to a thickness of 0.5
μm were deposited. The platinum deposits were examined by AES sputter
profiling and by electron microprobe analysis. No solution contaminants such
as oxygen or chlorine were found. Deposits of Pt and Au on undoped InP
exhibited ohmic behavior. In the initial phase of metal deposition, a
photothermal chemical reaction between the InP surface and the metal salt
seems to be involved. In the interfacial layer, PtP_2, NiP or similar
compounds may be formed. Further decomposition seems to occur via pyrolysis
at the metal-liquid interface. Attempts to deposit Pt on (111) Si were
unsuccessful. An understanding of the deposition process clearly requires
further investigations.

Electrochemical laser-enhanced plating has been studied in great detail by
VON GUTFELD [5.32,33]. Some of the experimental details have already been

outlined in Sect.4.1. The most extensively studied materials were Au, Cu and Ni, which were deposited on a variety of substrates. The substrates were glass and c-Al$_2$O$_3$, both covered with 0.1 μm thick films of either Au, Cu, Ni, Mo or W, and, additionally, metals such as Ni and Ni-plated Be-Cu (Table 5.1). The plating mechanism was investigated by illuminating the metallized glass surface from two directions: either from the front through the solution, or from the back through the optically transparent glass. In the latter illumination geometry, no photons reach the electrolyte, because the light is absorbed within 30 to 200 Å of the 0.1 μm thick metal film. Interestingly, the deposition rate was found to be equal for both geometries of illumination. Hence, nonthermal photochemical processes at the liquid-solid interface cannot play an important role. On the other hand, this result is expected for a thermally activated reaction, because the laser-induced temperature distribution within the metal surface is approximately the same for both illumination geometries. Further support for the thermal character of the deposition process was obtained from the comparison of plating results for premetallized c-Al$_2$O$_3$ (κ_D = 20 W/mK) and glass substrates. Under otherwise identical experimental conditions, the depositon rates on c-Al$_2$O$_3$ substrates were found to be a factor of 10 smaller than on glass. This can be directly explained by the decrease in center temperature (at constant laser irradiance) with increasing thermal conductivity of the substrate (Sect.2.1.1). The temperature change influences the charge and mass transfer rates within the electrochemical system (with and without an external EMF). The mechanism has already been described in Sect.3.3.

Detailed investigations into electrochemical Au plating have revealed that dense, small-grained, crack-free and uniform deposits of good adhesion are formed at elevated ambient temperatures and at a high density of gold concentration in the electrolyte. The operating potential was of similar importance; this should not exceed the mass-transport limit. Near this limit, Au of good morphology was deposited over areas 500 μm in diameter with rates up to 1 μm/s. Direct writing of Cu lines on premetallized glass substrates was possible with minimal widths of about 2 μm.

A new technique that improves the mass-transport limitation and thereby increases the deposition rate is laser-enhanced jet-plating (Sect.4.1). This technique allows high-quality, rapid, localized plating. The electrochemical and hydrodynamic parameters determining the mechanical and metallurgical properties of the deposits have been investigated, in particular for the model substance Au ⌊5.34,36,222⌋. For Au, plating rates of up to 12 μm/s have been achieved. The surface smoothness of the Au films increases with increasing laser power density. Simultaneously, the nodularity decreases and the voids even disappear. The Knoop hardness of the films ranged from 20 to 90 kg/mm^2. These values are characteristic for soft gold.

Electroless plating has been demonstrated for Ni and premetallized glass substrates. Deposition rates up to 0.1 μm/s have been achieved, while the background plating rates were about $5 \cdot 10^{-4}$ μm/s. Thermobattery or laser-enhanced exchange plating has been studied for Au and Cu on premetallized glass and phenolic resin paper substrates and on bulk metals [5.32,33,44]. The deposition rate for Cu spots was found to decrease with increasing thickness of the metal film covering the glass. This is in agreement with the thermal model for the plating process.

5.2.4 Solid-Phase Deposition

In this section we shall concentrate on the deposition of materials from solid films, such as metallopolymers or organosilicates, which have been produced by the spun-on technique. In this area, only a few investigations have so far been performed.

For solid-phase deposition of metals, one of the most interesting classes of compounds seems to be metallopolymers that are stable at room temperature. These metal inks can be spun onto substrates and baked to remove excess trapped solvent. Laser irradiation can decompose these films resulting in the formation of metallic structures. The first detailed experiments were performed by GROSS et al. [5.45-47,115] on the deposition of Au and Pd on SiO$_2$ and Si substrates by means of Ar$^+$ laser irradiation. JAN and ALLEN [5.200] have performed similar experiments with cw CO$_2$ laser radiation. For the Au compound, differential thermal analysis and thermogravimetry have shown that decomposition in air proceeds in two steps: a weakly endothermic stage at about 520 K, which is accompanied by substantial weight loss, and an extremely exothermic reaction step at 620-675 K [5.45-47]. From studies on model Au compounds it has been proposed that the weakly endothermic step corresponds to decomposition of the Au-containing compound in the ink, followed by volatilization of the organic constituent, while the highly exothermic step involves pyrolysis of a carrier polymer used to impart desirable rheological properties to the mixtures. In laser direct writing of Au structures, the samples were afterwards developed in CH$_2$Cl$_2$ in order to remove the unreacted metallopolymer film. Figure 5.16 shows the average line width of Au stripes as a function of laser power for two scanning velocities. The 514.5 nm Ar$^+$ laser line was used in these experiments. The width of lines

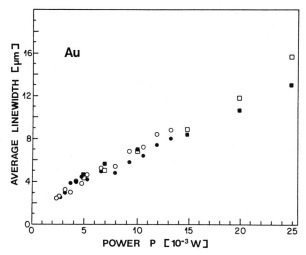

Fig.5.16. Average width of Au lines as a function of 515 nm Ar⁺ laser power.
●, ■: v_s = 36 μm/s; o, □: v_s = 206 μm/s. Metallopolymer film thickness 1.7 ±
0.1 μm after pre-baking. The two symbols for the same scan speed designate
data from different samples (after ⌊5.45⌋)

increases approximately linearly with a small intercept at about 6-10 mW. It
should be noted, however, that lines produced at powers < 10 mW were
incompletely reacted and did not survive rinsing in a mixture of aqueous
HNO_3, H_2CrO_4 and H_2SO_4 (5 minutes at 300 K). The rinsing also removed regions
of lines having a high C and S content ⌊5.45-47⌋. With increasing thickness
of the metallopolymer film the average linewidth was found to increase for
all laser scan speeds and laser powers. Complete transformation was achieved
in films with an initial thickness of 1.7 μm. The electrical resistivity of
the Au lines, normalized to the bulk value, is shown in Fig.5.17 as a
function of laser power. For P > 25 mW, the resistivity is about five times
that of evaporated Au films. The large error bars in the low power region
originate from the occurrence of a periodic superstructure. The formation of
periodic structures was also observed for other systems and it will be
discussed in Sect.5.2.5.

Another example of solid-phase deposition is pyrolytic direct writing of
SiO_2 patterns from spun-on films of organosilicate material in an organic-
based solvent with the general formula $Si(OR)_x(OH)_{4-x}$. KRCHNAVEK et al.
⌊5.171⌋ performed the first experiments with 514.5 nm Ar⁺ laser radiation and
Si-wafer substrates. Smooth, continuous stripes were drawn with a scanning
speed of 100 μm/s. Linewidths as small as 1 μm were obtained. The film
thicknesses were typically 0.5 μm but could be varied by varying the spun-on

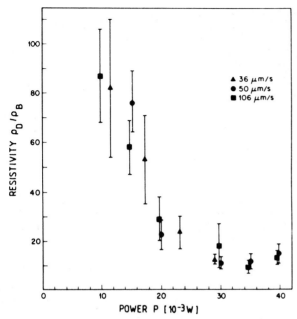

Fig.5.17. Resistivity of Au lines produced as in Fig.5.16 normalized to the bulk value for Au (2.44 Ω cm) (after [5.47])

speed and/or the viscosity of the organosilicate solution. Material not exposed to the laser beam was rinsed off with methanol. The laser-induced reaction probably consists of an initial elimination of the organic liquid and a subsequent release of OH groups. The quality of the SiO_2 films was at least as good as that obtained in thermally cured spun-on glass. The breakdown strength was about $1.6 \cdot 10^6$ V/cm.

5.2.5 Periodic Structures

In pyrolytic direct writing, various different types of periodic structures are observed in a large number of systems for certain ranges of processing parameters.

One type of periodic surface structure, the so-called ripples, which have been observed in laser-deposited thin films, arises from interference effects between the incident and scattered laser fields. Figure 5.18 shows examples for the case of Si deposited from SiH_4 on Si wafer substrates by means of 488 nm Ar^+ laser radiation [5.48,127,128]. These ripples show the characteristic features that have long been observed on many semiinfinite substrates, mainly after pulsed laser irradiation in a *nonreactive* atmosphere [5.202-204]. The

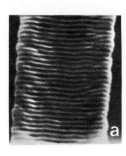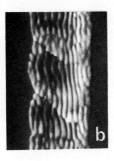

Fig.5.18a,b. Rippling on Si stripes deposited from SiH$_4$ on Si wafers by means of 488 nm Ar$^+$ laser radiation. The electric vector of the radiation was parallel (a) and perpendicular (b) to the direction of scanning (after ⌊5.48,128⌋)

ripples are characterized by some experimentally well-proved features: at normal incidence, the ripple spacing is approximately equal to the wavelength of the laser light, and the grating vector is polarized parallel to the incident electric field vector **F** (see Fig.5.18). These ripples will not be discussed any further in this section.

Another type of periodic structure having periods much longer than the ripples, has been observed in various different systems ⌊5.45-48,83,125,127, 128,201,223⌋. While the physical microscopic mechanisms are quite different for the various systems so far investigated, these structures seem to be a general phenomenon in laser-induced pyrolytic processing. Some examples will be discussed in the following.

Figures 5.19-20 show the new types of periodic structures for the case of direct writing based on pyrolytic LCVD, and for different systems and processing parameters. As far as is understood at present, these periodic structures are *not* related to the ripples mentioned above. No simple relation exists between the spatial period and the wavelength and the polarization of the laser light. The spacing of the periodic structures shown in the figures, i.e. the distance between the minima in the thickness or the width of the stripes, depends on the laser power, the scanning velocity, and the pressure of the reactant gas. These dependencies have been studied in detail for the deposition of Si on glass substrates covered with a thin layer of a-Si.

Figure 5.21 shows the spacing as a function of the incident laser power for a scanning velocity of v_s = 50 μm/s and a SiH$_4$ pressure of 200 mbar. While uniform stripes are observed for laser powers around 20 to 40 mW (case a in Fig.5.19, indicated by the crosses in Fig.5.21) periodic structures occur above and below this range of laser powers (the structure formation at lower laser powers seems to be closely related to the a-Si/glass substrate and will be discussed in ⌊5.224⌋). It becomes evident from Fig.5.21, however, that these limits in laser power are not clear-cut, as overlapping regions exist where stripes are either uniform or show pronounced periodic

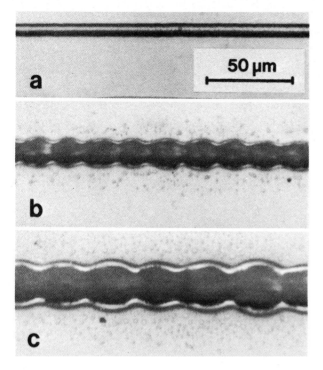

Fig.5.19a,b,c. Stripes of poly-Si deposited from SiH_4 by means of 647 nm Kr^+ laser radiation. The substrate is glass covered with 1200 Å a-Si. $2w_0$ = 7 μm, v_s = 50 μm/s, $p(SiH_4)$ = 200 mbar. (a) P = 34 mW; (b) P = 81 mW; (c) P = 130 mW (after [5.125])

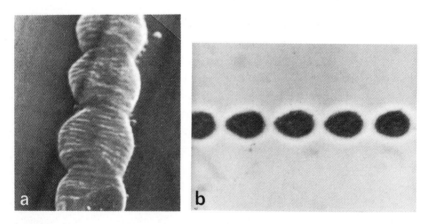

Fig.5.20a,b. Examples of periodic structures. (a) Si deposited on Si wafers (after [5.48,128]). (b) Ni deposited on 1000 Å a-Si/glass substrates (after [5.201])

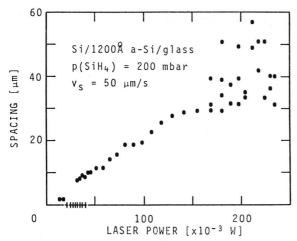

Fig.5.21.
Spacing of the periodic structures as a function of the laser power. + symbols indicate region of uniform deposition of stripes (after ⌊5.125⌋)

structures. For laser powers 40 mW $\lesssim$ P $\lesssim$ 150 mW, the spacing is well-defined and increases approximately linearly with laser power. For laser powers in excess of about 150 mW another instability is observed. In this regime a set of various different spacings, including the appearance of uniform structureless stripes, is observed for any given set of parameters. Figures 5.22a,b show the minimum (open symbols) and maximum (full symbols) widths of stripes as a function of the laser power for two different ratios $p(SiH_4)/v_s$. It can be seen that the amplitudes of the oscillations increase with increasing laser powers and ratios $p(SiH_4)/v_s$.

The origin of the structure formation remains speculative. First of all, the uniformity, fine reproducibility, and the measured dependencies on experimental parameters lead us to believe that the structure formation is not related to remaining fluctuations in experimental parameters - even in the light of the fact that, for example, minor changes in laser power (which was stabilized to better than 0.2%) and thereby in local temperature can significantly affect the deposition rate within the kinetically controlled region (see, for example, Fig.5.7). Furthermore, one can exclude any explanations that are based on the latent heat released by the SiH_4 decomposition and the concomitant formation of crystalline Si. For typical volumes of deposited material (laser power of 100 mW, SiH_4 pressure of 300 mbar, chemical heat release of 5 kcal/mole ⌊5.205⌋), one obtains a rate of heat release of the order of 10^{-6} W. This is negligible compared to the power absorbed from the laser beam, which amounts to about 50 mW (for a

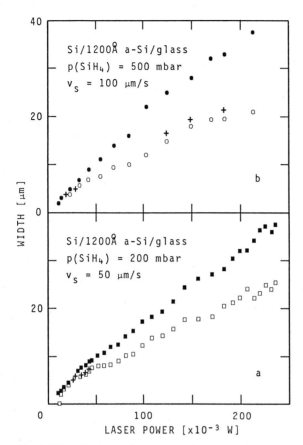

Fig.5.22. Minimum (empty symbols) and maximum (full symbols) widths of stripes as a function of laser power. (a) v_S = 50 µm/s, $p(SiH_4)$ = 200 mbar. (b) v_S = 100 µm/s, $p(SiH_4)$ = 500 mbar. Crosses indicate uniform stripes (after ⌊5.125⌋)

typical reflectivity of 40% - 50%, and a penetration depth below 1 µm for the relevant wavelengths and temperatures), unless one assumes that virtually all deposition takes place during single bursts that together last less than 10^{-4} of the dwell time of the laser beam. The rate of deposition of Si at a given site per unit of SiH_4 pressure is determined by the temperature at that site. In the case of a temperature-independent thermal conductivity κ and reflectivity R, and a high absorption coefficient α, this temperature distribution has been shown in Fig.2.14 for similar, but not identical geometries. From this figure, a qualitative explanation of the periodic structure for continuous stripes (rather than for separated single islands) is possible: Since the thermal conductivity of Si is at least a factor of 30 higher than that of the glass substrate, the heat flow via the stripe will soon dominate over the direct flow from the zone of deposition into the substrate. Thus, the surface temperature, and thereby also the temperature at the edge of the stripe (see dashed arrow on left hand side of Fig.2.14), will fall rather rapidly as the cross section of the stripe increases. The decrease in temperature leads to a reduction in the deposition rate, and hence to a decrease in the cross section of the stripe. The decrease in cross section, however, leads to an increase in local temperature (see dashed arrow on right hand side of Fig.2.14) etc. Due to the finite extent of the zone of deposition, this feedback mechanism is not instantaneous. The period of oscillations should decrease with increasing strength of the feedback.

The observation of even isolated islands of deposits could be tentatively understood in the following way: the laser light absorbed within the a-Si layer induces a temperature distribution with a center temperature barely exceeding the threshold for deposition. This results in the growth of a spot and a concomitant decrease in temperature. Once deposition has ceased it cannot start again until the overlap of the laser focus with the spot-like heat sink has decreased sufficiently for the threshold temperature for deposition to be attained once more.

The mechanisms outlined above can also explain the structures observed in Ni deposits on glass (Fig.5.20b), but not those seen in Si deposits on Si wafers (Fig.5.20a). In the latter case the thermal conductivity of the deposit is somewhat lower than that of the substrate, due to the higher temperature of the deposit $\lfloor$see (2.11)$\rfloor$. Thus, the surface temperature may be expected to increase with increasing thickness of the stripes. This continues until the surface reaches the melting point. At that point the reflectivity suddenly increases, leading to a lowering of the rate of energy absorption from the laser beam. As a consequence, the temperature and thereby the reflectivity will decrease. This results in an increase in absorbed laser power until the surface melts again.

Another type of periodic structure was recently observed by JACKMAN et al. $\lfloor 5.83 \rfloor$ during Ar^+ laser-induced deposition of Fe on SiO_2-glass substrates. Here, the structure formation seems to be related to an oscillation between the endothermic decomposition of $Fe(CO)_5$ and an exothermic solid state reaction between the metal deposit and the SiO_2 substrate. The periodic segregation into Fe and Fe_2SiO_4 phases along the stripe was confirmed by scanning Auger spectroscopy. The spacings were, typically, around 50 μm.

Periodic structures with spatial periods larger than the laser wavelength were also reported to occur in experiments on laser-induced chemical decomposition of metallopolymer films. FISANICK et al. $\lfloor 5.45-47,115 \rfloor$

115

interpret the periodic structures as a chemical analog of explosive crystallization ⌊5.206,207⌋, i.e. as a process driven by the release of chemical latent heat. On the other hand, the heat flow via the Au or Pd films produced by the reactions studied in ⌊5.45-47,115⌋ appears to be at least comparable to the heat flow directly into the substrate. Thus, the mechanism proposed above may well also play a role in these systems.

Periodic structures have also been observed during rapid-scan synthesis of Ge_xSe_{1-x} from sandwich films of Ge and Se ⌊5.223⌋. The formation of these structures has been explained in terms of an explosive reaction based on the release of latent heat during synthesis.

The preceding examples have shown that in pyrolytic direct writing of microstructures the local temperature, and thereby the reaction rate, may oscillate, even under stationary conditions, when the process parameters, such as the laser power, scanning velocity, physical properties of the substrate etc., are held constant. Such oscillations may originate from the nonlinear heat transport, the explosive release of heat in a strongly exothermic reaction, or from nonlinearities in the reflection and absorption of the material etc. The oscillations may decrease the ultimate resolution of structures obtainable in direct writing below the value expected for a stationary temperature distribution on the basis of the exponential dependence of the reaction rate on temperature (Sect.3.5). On the other hand, the periodic structures can be suppressed for certain ranges of process parameters, which must be determined experimentally for every system of interest. Clearly, further investigations of both an experimental and a theoretical nature are necessary in order to illuminate the various mechanisms present in different systems, and their importance in direct writing of microstructures.

5.3 Extended Thin Films

Extended thin films of metals, semiconductors and insulators have been deposited with laser beams at both perpendicular and parallel incidence. The different irradiation geometries, focusing possibilities and most commonly used lasers have been described in Sects.4.1 and 4.2.

For perpendicular incidence the situation is very similar to that discussed in Sect.5.2 for pyrolytic and photolytic deposition of microstructures. The main difference is that gas- or liquid-phase transport limitations become important at lower (thickness) deposition rates

(Sect.3.2). For the gas pressures typically used in LCVD, transport will be determined by two-dimensional diffusion if a tight line focus is used (Fig.4.2b), and by one-dimensional diffusion if the laser beam is unfocused or defocused.

For parallel incidence, decomposition can be based either on gas-phase heating (Sect.2.1) or on selective excitations of electronic or vibrational transitions (Sect.2.2). Clearly, excitation or dissociation of species will take place within the total volume of the laser beam.

Laser-induced gas-phase heating (parallel incidence) and laser-induced photolysis (perpendicular or parallel incidence) allow the deposition of thin films at low or moderate substrate temperatures. This is one of the most important advantages over conventional CVD, where uniform substrate temperatures of, typically, 500 - 2000 K are required. Plasma CVD (PCVD), which also allows deposition at lower substrate temperatures, is inherently associated with overall vacuum ultraviolet irradiation and heavy ion bombardment that may damage the substrate and simultaneously result in impurities being incorporated into it.

Therefore, LCP offers the unique possibility of depositing thin films onto substrates which are sensitive to elevated temperatures and/or to particle bombardment. Examples are prefabricated Si wafers, compound semiconductors and polymer foils.

In the following, pyrolytic and photolytic deposition of extended thin films will be discussed separately for metals, semiconductors and insulators. Thin film production by reactive and nonreactive laser sputtering and laser evaporation will be briefly mentioned.

5.3.1 Metals

Thin extended films of metals have been deposited mainly from the corresponding alkyls, carbonyls and fluorides. In most cases investigated, decomposition of these molecules was based on gas-phase photolysis with excimer laser or harmonics of Nd:YAG laser light. Generally, the films deposited at normal incidence have much better mechanical and electrical properties than those deposited at parallel incidence. This difference in physical properties is related to the laser-induced substrate heating which occurs at normal incidence only. However, good quality films can also be produced at parallel incidence when the substrate is preheated. This latter technique allows for proper control of the substrate temperature.

·Detailed investigations on large-area metal film deposition have been performed by DEUTSCH and RATHMAN ⌊5.26,180⌋, in particular for W. ArF excimer laser radiation at parallel incidence has been employed for photolytic decomposition of WF_6 in H_2. Silicon wafers and 4600 Å SiO_2/Si served as substrates. Good adherence and low electrical resistivity (about twice the bulk value) of films was achieved only when the substrate was heated to about 700 K. While this temperature would seem to be rather high, it should be noted that standard CVD of W on SiO_2 substrates is typically performed at about 1000 K. The excellent conformal coverage and surface morphology of laser-deposited W films can be seen from the scanning electron micrographs

Fig.5.23. Scanning electron micrographs showing a side and top view of a W film over SiO_2 steps on a Si substrate. Deposition was performed with ArF laser radiation from a mixture of WF_6 and H_2. T_S = 560 K (after ⌊5.26,180⌋)

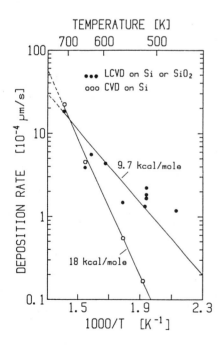

Fig.5.24. Arrhenius plot for the (thickness) deposition rate of W films. The precursor was WF_6 diluted in H_2. •: Deposited by ArF laser radiation on Si and SiO_2/Si substrates (after ⌊5.26,180⌋); o: Deposited by standard CVD on Si

shown in Fig.5.23. Figure 5.24 shows an Arrhenius plot for the deposition rate. For comparison, corresponding results obtained by standard CVD of W on Si are included. The figure shows that LCVD allows deposition of W on SiO_2 at temperatures where the rates in standard CVD are negligible, and also significantly enhances the rates for Si substrates at temperatures < 600 K. Another important feature seen in the figure is the strong decrease in the apparent chemical activation energy. This clearly shows that laser light irradiation changes the path of chemical reactions involved in the deposition process. It seems remarkable that such changes in the activation energy have not been observed in pyrolytic LCVD.

The following reaction mechanism was suggested: In the initial step the 193 nm ArF excimer laser radiation produces WF_n (n = 1-5) radicals. The F atoms released can react with H_2 in a strongly exothermic reaction according to

$$H_2 + F \longrightarrow HF + H \, . \tag{5.5}$$

The atomic H can then react with WF_6 and WF_n radicals via a complex series of reactions that result in the deposition of W and in gaseous HF as a reaction product. Such a reaction mechanism would at least qualitatively explain the strong change in activation energy with respect to standard CVD, where the rate-limiting step in the kinetically controlled region is probably determined by the dissociation of adsorbed H_2 molecules.

The laser-deposited W films are of high purity with a concentration of F < 1%. The sharp decrease in resistivities that was observed with increasing substrate temperatures can be explained by a structural change in the W films (see also Sect.5.4). It is well known that W exhibits a small-grain metastable β-W phase which transforms into a large-grain stable α-W phase at temperatures between about 600 and 950 K.

Extended thin films of Al [5.5,6], and Cr, Mo, W [5.6,68,69] have been deposited mainly with ArF and KrF lasers from $Al_2(CH_3)_6$ and from $Me(CO)_6$ (Me = Cr,Mo,W), respectively (Table 5.1). The substrate materials were mainly Pyrex, SiO_2 and Si. With perpendicular incidence, bright silvery metallic films of reasonable quality were obtained. Most of the films showed good adhesion to the substrates (> $5.5 \cdot 10^8$ dynes/cm^2 for SiO_2 substrates). The tensile stresses were between about 10^9 and $7 \cdot 10^9$ dynes/cm^2. Films of Cr and Mo deposited at 300 K had a tendency to peel when exposed to air. This could be avoided by heating the substrate to about 420 K either during deposition or prior to removal from the cell. Except for Al, the film resistivities were about 20 times higher than the corresponding bulk values (Table 5.1). These high film resistivities are probably due to C and C_xO_y impurities (see also Sect.5.4). Parallel laser beam irradiation led to grey or black particulate films with even higher carbon content.

Production of thin extended metal films should also be possible from the solid-phase, for example by transforming spun-on metallopolymers (Sect.5.2.4) over large areas, by using an optical configuration such as shown in Fig.4.2b.

5.3.2 Semiconductors

Thin films of semiconductors have been deposited on various substrates by laser irradiation at normal and parallel incidence (Sect.4.1). Mainly cw and pulsed CO_2 lasers, frequency-doubled Nd:YAG lasers and excimer lasers have been used in these experiments. The materials studied in most detail were Si, Ge and some compound semiconductors. The substrates used were mainly SiO_2, different types of glasses with complex composition and c-Si (Table 5.1).

a) Silicon, Germanium

By far the major part of the work has concentrated on thin films of amorphous hydrogenated Si (a-Si:H) deposited mainly by means of cw or pulsed CO_2 lasers. The most commonly used parent molecules are SiH_4 and Si_2H_6, which are often diluted with different buffer gases.

The enormous interest in a-Si:H arises from the fact that this material has become a leading candidate for low cost solar cell [5.208-212] and thin film transistor [5.210] applications. Hydrogen incorporation is necessary in a-Si, because it saturates Si dangling bonds and relieves strains, resulting in a reduced defect level and the ability to modulate the Fermi level by substitutional doping. Currently, a-Si:H is mainly produced by glow-discharge decomposition of SiH_4 or by reactive sputtering in argon-hydrogen mixtures [5.208-212]. Structurally superior amorphous films are produced by standard CVD. However, because CVD requires substrate temperatures of at least 900 K in order to obtain reasonable deposition rates, the films contain an insufficient amount of hydrogen (< 1 at. %) to achieve good electronic properties. Therefore, production of a-Si:H films by LCVD, which yields reasonable deposition rates at lower substrate temperatures, seems to be a promising alternative to the techniques currently employed.

CO_2 Laser-Induced Deposition

For CO_2 laser irradiation at normal incidence, deposition of Si on SiO_2 is dominated by pyrolysis at the gas-solid interface (SiO_2 strongly absorbs CO_2

laser light). Nevertheless, a wavelength dependence of the deposition rate that correlates with the absorption of gaseous SiH_4 was observed. This indicates that gas-phase heating is important even at normal incidence. The laser-deposited films were about 1 cm in diameter. Maximum deposition rates were about $2 \cdot 10^{-3}$ µm/s with SiH_4, and about 10 times faster with Si_2H_6 [5.149,150]. For Si_2H_6 the laser-induced threshold temperature for deposition was about 600 K.

Production of high-quality a-Si:H films with parallel incidence of the CO_2 laser light requires uniform substrate heating to, typically, 500 - 700 K. Deposition rates are much smaller than for normal incidence, even when significantly higher laser powers are used. For SiH_4, for example, typical values are $2 \cdot 10^{-4}$-$5 \cdot 10^{-4}$ µm/s. However, the uniformity of films is superior to those produced at normal incidence. Films with excellent adherence were deposited over areas up to 80 cm^2 [5.135,151,153,154,157,158,161].

Decomposition of SiH_4 seems to occur in two steps [5.213-216]. In the first step, which is considered as the rate-limiting process, SiH_2 radicals and H_2 are produced by gas-phase heating (Chap.2). This means that SiH_4 molecules are first vibrationally excited in a single-photon absorption process (indicated by *) and then transfer their energy via collisions. This first step can be schematically described by

$$SiH_4(g) + h\nu \longrightarrow SiH_4^* \longrightarrow \ldots \longrightarrow SiH_2(g) + H_2(g) \ . \tag{5.6}$$

This initial step is especially effective if the CO_2 laser frequency matches a strong vibrational transition of the SiH_4 molecule. This has been demonstrated by tuning the CO_2 laser wavelength - at constant laser power and under otherwise identical experimental conditions - from the 10.59 µm P(20) laser line, which is strongly absorbed by the ν_4-mode of the SiH_4 molecule, to the weakly absorbed 10.22 µm R(24) laser line. The difference in the laser-induced gas-phase temperatures for these two wavelengths, measured by N_2 CARS thermometry, was about 240 K [5.149,150]. The second step depends on the silane gas pressure and the laser fluence. At high pressures and/or laser fluences, SiH_2 molecules may react homogeneously with other SiH_2 or SiH_4 molecules to produce particles within the gas phase [5.216]. At lower pressures and laser fluences, diffusion of SiH_2 molecules into the substrate dominates and a thin film of a-Si:H will grow, following the reaction

$$SiH_2(g) \longrightarrow SiH_y(s) + (1-y/2) \ H_2(g) \ , \tag{5.7}$$

where y denotes the hydrogen concentration within the film.

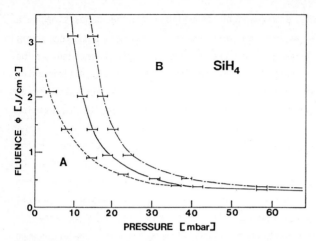

Fig.5.25. Region A shows the homogeneous decomposition threshold for controlled deposition of a-Si:H as a function of the SiH_4 pressure. Solid curve: pure SiH_4. Dash-dotted curve: SiH_4 + 40 mbar N_2. Dashed curve: $\overline{SiH_4}$ + 27 mbar SiH_2Cl_2. With parameters corresponding to those in region B, formation of powdery polycrystalline Si is observed (after [5.160])

The influence of gaseous admixtures on the absorption and homogeneous threshold for decomposition of SiH_4 under pulsed CO_2 laser irradiation has been investigated by PAULEAU et al. [5.139,140,160,161]. For N_2 buffer gas, the optical absorption increases with increasing N_2 partial pressure, mainly due to collisional deactivation of vibrationally excited SiH_4 molecules (Sect.2.2.2), while the Si deposition rate simultaneously decreases. On the other hand, as shown in Fig.5.25, admixture of N_2 to SiH_4 hinders homogeneous decomposition and the formation of powdery polycrystalline Si (region B). More regular and adherent a-Si:H films (region A) are thereby obtained over a wider range of SiH_4 partial pressures. Admixture of SiH_2Cl_2, however, decreases the threshold for homogeneous decomposition and thereby favors the formation of powdery silicon. Deposition was mainly performed on SiO_2 substrates that were homogeneously heated to about 620 K.

The deposition of B and P doped a-Si:H films by CO_2 laser irradiation at parallel incidence has been investigated by BILENCHI et al. [5.152] and by BRANZ et al. [5.156]. The precursors were SiH_4 diluted in Ar together with B_2H_6 and PH_3 for B and P doping, respectively. The substrates (SiO_2,Si) were uniformly heated to, typically, 500 - 700 K. The growth rates were $2 \cdot 10^{-4}$ - $5 \cdot 10^{-4}$ $\mu m/s$.

The composition, morphology and electrical properties of CO_2 laser-deposited a-Si:H films have been analyzed in great detail [5.135,151,153,154,156-158,213]. The main techniques employed were: X-ray diffraction, UV, visible and IR spectroscopy, hydrogen effusion and electron spin resonance (ESR). The magnitude of the unpaired spin concentration (this is a measure of the concentration of dangling bonds, which should be as low as possible for films of good quality) in the LCVD films was about 400 times lower than in conventional CVD films and only 5 times higher than in the best glow-discharge and HOMOCVD films ⌊5.214⌋. In other words, the quality of the laser-deposited a-Si:H films already approaches that of the best films produced by conventional PCVD techniques. This is all the more encouraging when one considers the enormous difference in the amount of effort which has been put into each of these techniques so far.

Excimer Laser-Induced Deposition

Amorphous and polycrystalline films of Si and Ge have also been deposited from SiH_4, Si_2H_6 and GeH_4 with excimer lasers. As we would expect from Sect.2.2.1, the growth rates strongly depend on the laser wavelength and intensity. The physical properties of these films and the amount of hydrogen incorporation have not yet been investigated in any great detail.

MURAHARA and TOYODA ⌊5.142⌋ have deposited films of a-Si:H from SiH_4 and Si_2H_6 on SiO_2 substrates by using perpendicular ArF and KrF irradiation. The laser beam was either unfocused or focused to a line as shown in Fig.4.2b. Decomposition of Si_2H_6 was very efficient with ArF but not with KrF laser light. With both lasers, decomposition of SiH_4 was only observed if the beam was focused. This difference in behavior is explained by the difference in absorption spectra. While 193 nm ArF laser radiation may photodissociate Si_2H_6 in a single-photon process, photodissociation of SiH_4 requires a multiphoton process at this wavelength (Sect.2.2.1).

EDEN et al. ⌊5.93-96⌋ have grown polycrystalline Si and Ge films from mixtures of SiH_4 + 90 mole % N_2 and GeH_4 + 95 mole % He, respectively. The Ge films were deposited on SiO_2, (100) NaCl and ($1\bar{1}02$) Al_2O_3 substrates mainly with KrF laser radiation at parallel incidence. The substrates were heated up to 500 K. The growth rates were between 10^{-4} and 10^{-5} μm/s. These rates are 10-100 times lower than those achieved at normal incidence. The average grain sizes of deposited films reached up to 0.5 μm. Germanium films doped with about 10^{20} Al atoms/cm^3 were obtained by admixture of $Al_2(CH_3)_6$. The homogeneous gas-phase decomposition mechanism of GeH_4 was verified from in situ measurements of the emission spectra. Several atomic Ge lines, most of which terminate on the $^1D_2$ metastable state, and all the transitions between 250 nm and 300 nm characteristic for neutral Ge were observed. Additionally, the A $^2\Delta \longrightarrow$ X $^2\Pi$ band of GeH was detected. From these observations, and the weak dependence of the deposition rate on substrate temperature, it was concluded that the transport of germanium to the substrate occurs mainly in atomic form ⌊5.95,96⌋. Clearly, a detailed understanding of the gas-phase photochemistry and kinetics of the deposition process requires further investigations. For possible applications, a detailed characterization of the films is indispensable.

b) Compound Semiconductors

DONNELLY et al. ⌊5.99-101⌋ have investigated photochemical LCVD of InP on SiO_2, GaAs, GaAlAs, (100) InP and Si substrates. Most of the experiments were performed with ArF laser light at perpendicular incidence. The precursor molecules were $(CH_3)_3InP(CH_3)_3$ ⌊this adduct decomposes into $In(CH_3)_3$ and $P(CH_3)_3$ above about 350 K⌋ and $P(CH_3)_3$. The added He and H_2 served as carrier and scavenger gases. Stoichiometric films of InP with thicknesses of up to more than 1 μm were deposited with rates of up to 0.4 Å/pulse. The films were characterized by SEM, TEM and RBS. The microstructure of films ranged from amorphous via polycrystalline to epitaxial, depending on the incident laser fluence and the substrate temperature, which was below 620 K. The best films, epitaxially grown on (100) InP substrates, were obtained with a laser fluence of $\phi \approx 0.1$ J/cm^2 at a substrate temperature of about 590 K. Under these conditions, the dark growth rate was zero or negligible. Note that this temperature is below the incongruent decomposition temperature of InP, which is about 620 K. It is remarkable to note that the laser power density on the substrate surface is a very critical parameter that affects crystallization, film and substrate damage, and impurity incorporation. In fact, laser surface irradiation has been found to be very efficient in suppressing contaminations of C and C_xH_y. This is in agreement with similar observations made for ArF laser-deposited Al films (see ⌊5.4⌋ and Sect.5.4). For properly deposited InP films, the C contamination was below detectibility by AES. In situ gas-phase fluorescence measurements allowed further insight into the photodecomposition mechanisms and also made it possible to monitor metal-organic precursor concentrations.

Laser-stimulated MOCVD of epitaxial (100) GaAs by means of frequency-doubled Nd:YAG laser light has been investigated by BENEKING and co-workers ⌊5.88,89⌋. The precursor molecules were $Ga(CH_3)_3$ and AsH_3 diluted in H_2. Figure 5.26 shows the growth rate of GaAs with and without laser light irradiation as a function of the uniform substrate temperature. It becomes evident that laser-assisted growth can be extended to much lower temperatures. The morphology of films strongly depends on laser fluence.

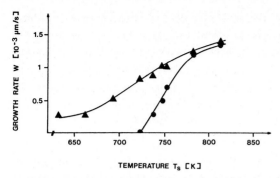

Fig.5.26. Growth rate of GaAs as a function of temperature. •: Conventional MOCVD; ▲: MOCVD enhanced by 532 nm pulsed Nd:YAG laser irradiation (after ⌊5.88⌋)

Smooth monocrystalline layers were obtained with a fluence of 120 mJ/cm^2. At a substrate temperature of 720 K, these layers were p-type ($3 \cdot 10^{17}$ cm^{-3}). Photoluminescence measurements have revealed this doping to be due to C incorporation. Mercury lamp assisted MOCVD of GaAs has been reported by PÜTZ et al. ⌊5.217⌋. IRVINE et al. ⌊5.97,98⌋ have investigated the Hg lamp photosensitized growth of HgTe.

c) Oxides

Zinc oxide, ZnO, which is an n-type semiconductor, has been deposited from $Zn(CH_3)_2$ with either NO_2 or N_2O as oxygen donor ⌊5.184⌋. Here, ArF and KrF excimer lasers were employed. For deposition rates of up to $5 \cdot 10^{-3}$ μm/s, the films were of good optical quality, low pinhole density, good adherence and good thickness uniformity (better than 5%) over areas of 10 cm^2. At higher deposition rates the adherence and quality of films deteriorated. The stoichiometry revealed by ESCA was 49% Zn and 51% O. The concentrations of impurities such as C and N were below 1%. By varying the ratio of partial pressures for $Zn(CH_3)_2$ and NO_2, the stoichiometry of films can easily be changed. Resistivities may then range from 10^3 to 10^{-1} Ωcm for 0.5 μm thick films. The simple control of the film stoichiometry is one of the advantages of LCVD over conventional techniques. The deposition of ZnO is of relevance for acoustoelectric and acoustooptic applications that are based on the large piezoelectric and optical coupling constants of this material. Additionally, its 3.3 eV band gap has attracted interest in using ZnO as a coating for solar cells, because it is less expensive than indium tin oxide (ITO).

Thin films of In_2O_3 have been deposited with ArF laser radiation at glancing (5^0) incidence ⌊5.100,101⌋. SiO_2, GaAs, and InP served as substrates. The parent molecules $(CH_3)_3$ $InP(CH_3)_3$ + $P(CH_3)_3$ + O_2 or H_2O were diluted in H_2 and He (see Sect.5.3.2b). Good-quality, nearly stoichiometric (In:O = 2 : 3.2 ± 0.3) films have been obtained at elevated substrate temperatures (around 600 K). No C or P impurities were detected with sputtering AES. Perpendicular irradiation under otherwise identical conditions resulted in metallic-looking films containing In and O in a ratio of 1 : 1 with no C or P.

d) Laser Sputtering and Evaporation

Besides the deposition from the gas phase, extended thin films may also be produced from the solid phase by laser-beam sputtering or evaporation ⌊5.218,225,226⌋. Investigations for Si have been performed by HANABUSA et al. ⌊5.167,219⌋. With a frequency-doubled Q-switched Nd:YAG laser, deposition rates of 10 to 100 μm/s were obtained. Similarly to ion-beam sputtering, laser sputtering can also be performed in a reactive atmosphere. For example, a-Si:H films have been deposited by laser-induced evaporation of Si in a hydrogen atmosphere ⌊5.163⌋.

Thin, high-quality films of ZnO have been produced by CO_2 laser evaporation of ZnO powder ⌊5.186⌋.

5.3.3 Insulators

This section deals with the deposition of extended insulating films by laser-induced gas-phase or solid-phase reactions. There seems to be no example of this using liquid-phase deposition. Direct surface modification such as oxidation, nitridation, etc. by laser-induced heating in an O_2 or N_2 atmosphere is discussed separately in Sect.6.1.

Laser-induced CVD of extended thin films of insulators such as SiO_2, TiO_2, Al_2O_3 or Si_3N_4 (Table 5.1) has mainly been demonstrated with ArF, KrF and CO_2 lasers. As described previously, laser irradiation can be performed at either normal or parallel incidence or by a combination of both (Fig.4.3).

a) SiO_2

Most of the investigations have concentrated on the deposition of amorphous SiO_2 (a-SiO_2; note that we frequently denote quartz glass simply by SiO_2, while crystalline quartz is denoted by c-SiO_2). Films of SiO_2 are an integral part of every semiconductor device, providing electrical insulation and passivation as well as masking layers for both the diffusion of dopants and pattern transfer in etching. Additionally, SiO_2 layers are often used for materials passivation against corrosion.

BOYER et al. ⌊5.6,28,29,164,166-168⌋ have deposited extended thin films of SiO_2 mainly with ArF laser radiation. The parent molecules consisted of a mixture of 1%-5% SiH_4, 89%-85% N_2O and 10% N_2. The total gas pressure was typically ⩽ 10 mbar. Mainly Si wafers were used as substrates.

For *parallel* incidence, deposition rates of 10^{-3}-$5\cdot10^{-3}$ μm/s (it should be noted that the upper value reported in ⌊5.167⌋ seems to be too high for good quality films) were obtained. The deposition rates were directly proportional to the gas pressure and the laser intensity, but independent of the substrate temperature, which was varied between 300 K and 900 K. The quality of films, however, did depend very sensitively on substrate temperature. For temperatures ⩾ 520 K, clear, stoichiometric, scratch-resistant and adherent films of good uniformity over areas of 1.5x7.5 cm^2 were obtained. For substrate temperatures < 470 K, the films were milky. The physical, electrical and chemical properties of the films as a function of substrate temperature were investigated in detail [5.6,28,29,166,167]. Surface states, and flat band and breakdown voltages were measured with polysilicon gate MOS capacitors and evaporated Al gate capacitors. Additionally, IR absorption, index of refraction, etch rate and pinhole density measurements were performed. Although N_2 and N_2O are the major gas constituents, sputtering Auger analysis revealed a nitrogen content of only 2%-4%. The pinhole densities (< 1/cm^2), dielectric breakdown voltage (> 9 MV/cm) and the step-coverage abilities were comparable to thermally grown native oxides. The main properties of SiO_2 films produced by LCVD, PCVD and EB processing are compared in Table 5.2.

Deposition of SiO_2 by *combined parallel and perpendicular* irradiation with ArF laser light has also been studied by BOYER et al. [5.164]. Apart from the changes in irradiation geometry, the experimental conditions were similar to those described above. The parallel beam creates a thin sheet of photofragments in a localized region just above the substrate with minimal interaction with the surface. The perpendicular beam creates transient heating of the SiO_2/Si interface. This dual beam configuration allows for in situ annealing and surface heating, as well as photostimulation of surface reactions. The deposition rates were increased by 10%-20%. Furthermore, a reduction in the chemical etch rate and an increase in the refractive index were observed. These observations were associated with the reduction of SiH bonds below the IR detectibility, and of SiOH bonds by over 50%. Good-quality films with deposition rates of about $1.5\cdot10^{-3}$ μm/s were obtained at substrate temperatures of about only 500 K.

127

Table 5.2. Physical and chemical properties of extended thin films of SiO_2 deposited by various techniques (after [5.6, 28, 29, 168]). The abbreviations employed are listed at the end of the book

	LCVD	PCVD	EBCVD
T_s [K]	400 – 700	650	500 – 700
p_{tot} [mbar]	10	1.4	0.35
Gas $N_2O/SiH_4/N_2$	80/1/40	33/1/10	75/1/75
W [μm/s]	$1.7 \cdot 10^{-3}$	$5 \cdot 10^{-4}$	$9 \cdot 10^{-4}$
CVD parameters	$\lambda = 193$ nm	450 kHz	4.7 kV
	100 Hz		12 mA
	10-ns pulse		
Adhesion [10^8 dyne/Cm^2]			
(1000 Å on Si)	>7	>7	>7
Pinholes/cm^2			
(1000 Å on Si)	1 – 100	25 – 100	100 – 700
(2000 Å on Si)	~ 1	~ 1	10 – 100
N (632.8 nm)	1.46	1.46	1.46
Stress on Si			
[10^9 dyne/cm^2]			
(compressive)	1.5	3.6	9.4
Breakdown voltage [MV/cm]			
(1000 Å on Si)	6 – 8	–	2 – 6
(2000 Å on Si)	–	10	–
ρ at 5 MV/cm [Ω cm]	$10^{13} - 10^{14}$	10^{16}	$10^{14} - 10^{16}$
Flat band voltage [V]	2 – 10	<0.2	0.5 – 3
ε_d at 1 MHz	3.9 – 4.6	4.6	3.5
	(thermal oxide = 3.9)		
Stoichiometry	SiO_2	$SiO_{1.94}N_{0.06}$	SiO_2
N [%]	<1	3	<1
C [%]	<2	<0.1	<2
Hydrogen bonding			
(2270 cm^{-1} SiH [%])	2.3	2	<
(3380 cm^{-1} H_2O [%])	~ 0.1	<0.001	<0.001
(3650 cm^{-1} OH [%])	0.6	0.002	<0.01
Etch rate [Å/s] in			
7:1 buffered HF			
(as deposited)	30	20	30 – 60
(after 60 min			
1220 K N_2 anneal)	~ 18	–	–

The detailed photochemistry of SiO_2 photodeposition is not yet well understood. Some remarks on the photochemistry of SiH_4 and N_2O have been made in Sect.2.2.1. The deposition kinetics is controlled by a competition between quenching and recombination of atomic oxygen, oxidation of silicon hydrides, creation of reactive nitric oxide species, and substrate reactions. From the low nitrogen concentration within the films, it was concluded that only the oxygen species $O(^1D,^3P)$ and O_2^* are important in the film growth kinetics ⌊5.166⌋.

Laser-induced formation of extended SiO_2 films from the solid phase has not yet been realized. However, by using high power lasers, it should be possible to transform large areas of spun-on organosilicates into SiO_2. The method has been described for microstructure fabrication in Sect.5.2.4.

b) TiO_2

TiO_2 has been deposited from a mixture of $TiCl_4$ + H_2 + CO_2 on quartz substrates by means of CO_2 laser radiation at normal incidence ⌊5.109,110,112⌋. At higher temperatures, H_2 and CO_2 combine to generate H_2O, which reacts with $TiCl_4$ within the laser-heated substrate area. The film thickness increases approximately linearly with laser-beam illumination time. Deposition rates of about 0.3 μm/s were achieved. The films were in general clear, had good adherence and relatively low pinhole density.

c) Al_2O_3

Al_2O_3 is increasingly considered as an attractive substitute for SiO_2, mainly in microelectronics. Additionally, the material has many other applications which also benefit from a low temperature, high deposition rate process (Chap.9).

SOLANKI et al. ⌊5.5,6,28,29⌋ have used ArF and KrF laser radiation at parallel incidence and a mixture of $Al_2(CH_3)_6$ and N_2O as precursors. The substrate materials used were glass, and wafers of Si and III - V compounds. By moving the substrate relative to the laser beam, films of uniform thickness (± 5%) across 7.6 cm (3 in.) wafers were deposited at rates of about $3 \cdot 10^{-3}$ μm/s. When the deposition rates were increased up to $2 \cdot 10^{-2}$ μm/s, the films were of bad adherence. For even higher rates, gray powdery films were obtained. Table 5.3 facilitates comparison of the physical and chemical properties of good-quality LCVD and RF sputtered films. The mechanical, optical and electrical properties of the LCVD films can be further improved by using combined parallel/perpendicular laser irradiation (⌊5.25-27⌋; see also Fig.4.3b and Sect.5.4).

Table 5.3. Physical and chemical properties of extended thin films of Al_2O_3 deposited by LCVD [5.6, 28, 29] and by RF sputtering [5.220]. The abbreviations employed are listed at the end of the book

	LCVD	RF sputtered
T_s [K]	400 – 700	400 – 700
p_{tot} [mbar]	1.3	$1.3 \cdot 10^{-2}$
W [µm/s]	$3 \cdot 10^{-3}$	$5 \cdot 10^{-4}$
CVD parameters	$\lambda = 193$ nm	
	100 Hz	
	10-ns pulse	
Adhesion [10^8 dyne/cm^2]		
(1000 Å on Si)	>6.5	Strongly adherent
Pinholes/cm^2	<1	31
	(1000 Å)	(2500 Å)
Stress [dyne/cm^2]	$<6 \cdot 10^9$ (tensile)	$2.8 \cdot 10^9$ (compressive)
n	1.63	1.66
ρ [Ω cm]	10^{11}	10^{12}
ε_d at 1 MHz	9.74	9.96
Etch rate in 10% HF [Å/min]	100	–
Stoichiometry	Al_2O_3	$Al_{2.1}O_3$
Impurities [%]	C<1	Ar ~ 5

A fuller understanding of the photochemistry of the Al_2O_3 photodeposition process must await further investigations. Previous experiments on the UV photodissociation of N_2O have shown that the primary reactive products are excited atomic oxygen and ground state N_2 (Sect.2.2.1). The Al_2O_3 seems to form in subsequent reactions between photodissociated Al fragments and atomic oxygen.

d) Si_3N_4

Si_3N_4 films have been deposited with ArF laser light at parallel, perpendicular or combined parallel/perpendicular incidence [5.6,25,28,29]. In the latter configuration, the energy density of the perpendicular beam was about 1/10 that of the parallel beam. The parent molecules were SiH_4, NH_3 and N_2 in a ratio 1:(1-10):40 [note that the ratio of SiH_4 to NH_3 is much higher than that for SiH_4 and N_2O used for the deposition of SiO_2; the reason is that for 193 nm radiation $\sigma_a(NH_3) \approx 10^3 \, \sigma_a(N_2O)$]. The total gas pressure was about 3 mbar. Deposition of Si_3N_4 was mainly performed on Si, SiO_2 covered Si wafers and ZnSe. The substrate temperatures were varied between 320 K and 900 K. The deposition rate increased linearly with laser power (indicative of a single-channel dissociation process) and with total gas pressure. Deposition rates of up to $1.2 \cdot 10^{-3}$ µm/s were achieved. The physical and chemical properties of the films have been studied by DEUTSCH et al. ⌊5.25⌋ and by EMERY et al. ⌊5.6,28,29⌋. The stoichiometry and the physical properties of LCVD, PCVD and CVD films may be comparable. That is, good-quality LCVD films

Table 5.4. Physical and chemical properties of extended thin films of Si_3N_4 deposited by various techniques (after [5.6, 28, 29, 168]). The abbreviations employed are listed at the end of the book

	LCVD	PCVD	EBCVD
T_s [K]	300 – 700	650	300 – 700
p_{tot} [mbar]	3	3	0.5
Gas $NH_3/SiH_4/N_2$	1/1/40	7/1/0	60/1/44
W [µm/s]	10^{-3}	$5 \cdot 10^{-4}$	$3 \cdot 10^{-4}$
CVD parameters	$\lambda = 193$ nm	450 kHz	4.2 kV
	100 Hz		25 mA
	10-ns pulse		
Adhesion [10^8 dyne/cm^2]			
(1000 Å on Si)	>6	>6	>5.5
Pinholes/cm^2			
(1000 Å on Si)	10 – 100	2 – 3	5 – 100
(2000 Å on Si)	<1	<1	–
n (632.8 nm)	1.85	2	1.85
Stress [dyne/cm^2]	$4 \cdot 10^9$ compr.	$4.7 \cdot 10^9$ compr.	
ρ_m [g/cm^3]	2.4	2.8	
Breakdown voltage [MV/cm]	2.5	4	6
ρ at 1 MV/cm [Ω cm]	$10^{14} – 10^{15}$	$10^{15} – 10^{16}$	$10^{12} – 10^{14}$
ε_d at 1 MHz	7.1	7	7.1
Stoichiometry	$Si_3N_{4.3}$	Si_3N_4	Si_3N_4
O [%]	5	<1	<0.1
C [%]	4	<1	<0.1
Hydrogen bonding			
SiH [%]	12	12 – 16	<0.1
NH [%]	11	2 – 7	8 – 10
Etch rate [Å/s]			
in 5:1 buffered HF	20 – 50	1.7	3 – 20

possess good adherence, low pinhole density, low compressive stress, and excellent step coverage. The dielectric properties of the LCVD films, however, are not yet satisfactory. Table 5.4 summarizes some typical properties of films.

5.4 Morphology and Physical Properties of Deposits

Throughout this chapter many remarks on the morphology and the physical properties of deposits in the form of microstructures and extended thin films have been made. In this section, we shall confine ourselves to summarizing some of the common features.

Most of the materials deposited by the various techniques were polycrystalline. Epitaxial and single-crystalline growth has been demonstrated in only a very few exceptional cases (see Table 5.1). For pyrolytically deposited films, the grain sizes increase with increasing laser power. This was studied in some detail for Ni ⌊5.50,105⌋. Photolytically deposited films show, in general, smaller grain sizes than pyrolytically deposited films. The adherence of films depends strongly on the substrate material, the foregoing cleaning procedure, and the laser power and wavelength. For example, good adherence was found for pyrolytically deposited Ni stripes on all substrate materials investigated, except for 1000 Å a-Si/glass substrates and laser powers > 20 mW. The adherence of the Ni stripes increased with decreasing laser wavelength. Photolytically deposited films show good adherence as well and generally pass the Scotch tape test.

The electrical properties of deposits were investigated for some systems and processing conditions. This is indicated by a ρ in Table 5.1. For pyrolytically deposited metal films the resistivity decreases with increasing laser power and is typically 1 to 5 times higher than the bulk resistivity. Photolytically deposited metal films produced without substrate heating have resistivities that may exceed the corresponding bulk values by a factor of 10-10^4. The large differences in electrical properties of pyrolytically and photolytically deposited films originate from the differences in grain sizes and, more importantly, the concentrations of impurities incorporated in the films. When films are being deposited from alkyls or carbonyls, the concentrations of C_xH_y or C_xO_y fragments, which strongly influence the electrical properties, decrease with increasing substrate temperature. This explains the superior electrical properties of pyrolytically deposited films. There are several possible ways of improving the morphology and physical

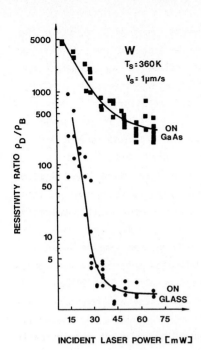

Fig.5.27. Resistivities of laser-deposited W stripes normalized to bulk values as a function of incident power of the 351-364 nm Ar$^+$ laser output. No buffer gas was used. The precursor gas pressure was $p(W(CO)_6) \approx 0.05$ mbar (after [5.103])

properties of photodeposited films. One possibility is the uniform heating of the substrate (indicated in Table 5.1 by HS) to a temperature where no dark reaction - or only a negligible one - takes place. Another possibility is the combined twin-beam or single-beam pyrolytic-photolytic technique that has been described in Sects.4.1 and 5.2.2. In this connection it may be illuminating to briefly discuss some resistivity measurements on metal stripes that have been deposited with near UV laser light.

Figure 5.27 shows, for the example of W and two different substrates, the resistivities of stripes as a function of the power of the 350-360 nm Ar$^+$ laser output [5.103]. The precursor was $W(CO)_6$ without any buffer gas. The most remarkable feature in the figure is the difference in resistivities of W stripes on GaAs and glass substrates. This behavior ultimately originates from the differences in the optical and thermal properties of these substrates. For example, the thermal conductivity of GaAs exceeds that of glass by a factor of 30 to 10, itself depending on the temperature. Therefore, the local laser-induced temperature at a certain laser power will be considerably lower on GaAs than on glass substrates (Sect.2.1.2). Consequently, the relative importance of pyrolytic and photolytic decomposition mechanisms will be quite different for these substrates. The lower laser-induced temperature on GaAs favors the growth of granular films

133

and, additionally, incorporation of photofragments, such as C, O, CO or CO_2, into the W films. This goes some way towards explaining the great differences in the resistivities. For W, however, the situation is somewhat more complicated. Here, the higher film resistivities measured on GaAs substrates may also originate, in part, from the different metallurgical phases of W that may form at different temperatures (Sect.5.3.1). Even in bulk material, the electrical conductivity of the α-phase, which is formed between 600 and 950 K, exceeds that of the β-phase by a factor of 100 to 300.

There seems to be another possible way of improving the physical properties of photodeposited films. HIGASHI and ROTHBERG [5.4] have recently shown that the CH_3 contamination within Al films photolytically deposited from $Al_2(CH_3)_6$, can be significantly reduced by subsequent ArF laser irradiation in vacuum. After irradiation, the films had similar electrical and optical properties to those deposited at elevated substrate temperatures (see [5.5] and Sect.5.3). These results seem of particular interest and should be further substantiated for other films deposited from alkyls and also from carbonyls. If this observation holds more generally, it would be a unique tool to improve the quality of both photolytically and pyrolytically deposited films. In the latter case this would be of particular relevance in high resolution pyrolytic patterning, which is achieved at low local laser-induced temperatures (Sect.3.5). It may be that the improvement of other laser-deposited materials, as e.g. Al_2O_3, by UV laser irradiation can be explained, at least in part, along similar lines (Sect.5.3.3).

6. Surface Modifications

This chapter deals with laser-induced surface oxidation, nitridation, reduction, metallization and doping. In this processing mode, atoms or molecules (including photofragments) of the adjacent medium either directly combine with atoms or molecules within the solid surface, or they simply diffuse into this surface. Another type of surface modification is the laser-induced depletion of a particular component of the surface, without appreciable etching or evaporation of the solid. In each of these treatments the physical and/or chemical properties of materials' surfaces will be modified. Surface modification, in general, requires (thermal and/or nonthermal) photoexcitation of the substrate, and is therefore performed mainly at normal incidence. Large-area surface modifications are performed with high power CO_2 lasers, excimer lasers, Nd:YAG or Nd:glass lasers, while for local modifications low power cw lasers such as Ar^+ or Kr^+ lasers may be employed.

6.1 Oxidation, Nitridation

Surface oxidation of metals and semiconductors in an oxygen-rich atmosphere is a well-known phenomenon. Clean surfaces of materials like Al, Nb, Si, etc. spontaneously react in air, even at room temperature, to form thin native oxide layers (typically 10-100 Å thick). This thermally activated process is self-terminating, because oxygen diffusion becomes less likely with increasing layer thickness. For many applications (local hardening, chemical passivation, electrical insulation, etc.), it is desirable to increase the thickness of native oxide layers, or to stimulate oxidation on material surfaces that do not spontaneously oxidize in an oxygen environment. A large number of techniques for producing oxide layers on solid surfaces have been investigated. The most commonly used such techniques are thermal oxidation by uniform substrate heating in an oxygen-rich atmosphere and plasma oxidation. Both techniques have their characteristic advantages and disadvantages, which are briefly discussed in Chap.9. A novel technique is laser-enhanced surface oxidation. It may again be based on thermal and/or nonthermal physical mechanisms. Laser-enhanced pyrolytic (thermochemical) oxidation can be understood along the lines of the preceding chapters: The absorbed laser light induces a surface temperature rise (Sect.2.1.1) and thereby enhances the diffusion and reaction of species within the irradiated area. The resulting oxide layers have a graded composition, very similar to that observed in films produced by uniform substrate heating. The situation becomes different at high laser irradiances, in the so-called regime of laser-pulsed plasma chemistry (LPPC). Here, the laser light initiates a breakdown within the ambient atmosphere at or near the solid surface. The breakdown threshold depends on the ambient atmosphere and the physical properties of the solid surface ⌊6.1,2⌋. For air-metal interfaces and CO_2 laser radiation, typical threshold intensities are of the order of 10^6 W/cm^2 (for homogeneous breakdown in air, i.e. far from interfaces, the value is about 10^3 times higher) ⌊6.107⌋. In LPPC the laser-generated plasma can initiate excited reactants, which can combine with the substrate atoms to form overlayers such as oxides. Oxide layers produced in this way are very thin and of similar composition to those produced by plasma oxidation. Another mechanism that has also been extensively studied is the surface oxidation of semiconductors by laser-induced band-gap excitation. Here, the photon energy, hv, must exceed the energy of the gap between the valence and the conduction band, E_g. Band-gap excitation can modify the adsorption of species at the surface and/or their migration into the surface, the reaction with surface atoms, etc.

Nitridation of solid surfaces is generating increasing interest as an alternative to oxidation. The techniques employed are similar to those for surface oxidation.

In the following we shall discuss laser-enhanced surface oxidation and nitridation separately for metals and semiconductors. Table 6.1 lists the various systems investigated and also gives an overview of the extensive literature. Note that laser-induced deposition of oxides and nitrides was outlined in Chap.5, and is not included in this section.

6.1.1 Metals

Detailed investigations on laser-enhanced thermochemical *oxidation* of metals have been performed by METEV et al. ⌊6.7-9⌋, URSU et al. ⌊6.10-15⌋ and WAUTELET et al. ⌊6.3-6⌋. While in numerous investigations the surface layer oxidized was only very thin in comparison with the thickness of the

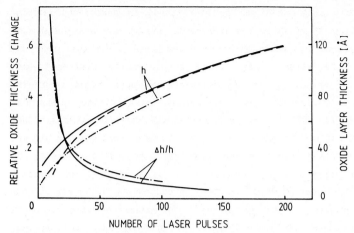

Fig.6.1. Thickness, h, and relative change in thickness, Δh/h, of Cr_2O_3 formed by pulsed Nd:glass laser irradiation of Cr on SiO_2. Full and dashed curves are experimental results. The dash-dotted curve is calculated (after [6.7])

substrate, several investigations performed on thin metal films (h ⩽ 0.1 μm) succeeded in bringing about oxidation throughout the whole thickness. In these latter cases the formation of stoichiometric compounds has often been observed. Clearly, these investigations could equally well be incorporated in Chap.7.

Figure 6.1 shows the dependence of the oxide layer thickness on the number of pulses of a Nd:glass laser for the example of Cr (full and dashed curves). It becomes evident that oxide growth is most efficient during the first several laser pulses. This initial regime of oxidation can be well described by the Cabrera-Mott oxidation theory (dash-dotted curves) [6.61,62]. With increasing oxide layer thickness the barrier for diffusion of species (mainly metal) increases and, as a consequence, the oxidation rate decreases.

Niobium films have been oxidized in 1000 mbar O_2 by XeCl excimer laser radiation [6.28,29]. The chemical composition investigated by XPS was found to be very similar to that of films produced by thermal oxidation.

Enhanced surface oxidation based on laser-pulsed plasma chemistry (LPPC), was first studied by MARKS et al. [6.28,29]. These investigations concentrated on the oxidation of Nb films (typically 130 μm thick) in an O_2 atmosphere by pulsed CO_2 laser radiation. Here, single-pulse laser-activated oxidation was observed to produce thicker films than multiple-pulse irradiation. For example, for a single pulse with a fluence of $\varphi = 0.75$ J/cm^2, the thickness of the native oxide consisting of $Nb_2O_{5-\delta}$ was increased by 18 Å, while 3 pulses each having a comparable fluence yielded a

137

net increase of only 11 Å. This can be understood, in part, by competing mechanisms: Oxidation by LPPC on the one hand, and oxide ablation due to absorption of CO_2 laser radiation within the oxide layer on the other. As revealed by XPS, the valence defect, δ, monotonically decreases with increasing layer thickness. A similar behavior has been found for films produced by plasma oxidation. However, for a given layer thickness, the δ is 3 to 5 times smaller for LPPC oxides ($0.02 \leqslant \delta \leqslant 0.04$) than for plasma oxides ($0.1 \leqslant \delta \leqslant 0.2$). In other words LPPC yields more complete oxidation. Furthermore, in comparison with laser-enhanced thermochemical or conventional oxidation, LPPC strongly reduces the formation of suboxides. The niobium oxide layers (18-40 Å for $0.24-0.79$ J/cm^2) produced by LPPC may become essential for fabrication of tunnel barriers in tunneling devices. It is remarkable to note that it is difficult to produce such well-defined dielectric layers with comparable thickness control and quality by the standard techniques.

Nitridation of metal surfaces was studied mainly for Ti and Zr by both pulsed and cw CO_2 laser irradiation. The ambient media were gaseous and liquid N_2, and air ⌊6.54-56,60⌋. As expected, the nitridation process is extremely sensitive to traces of oxygen within the nitrogen. The nitrified Ti and Zr surfaces were nonuniform with varying mixtures of TiN, TiN_xO_y, TiO_2 and ZrN, ZrN_xO_y, ZrO_2, respectively. The composition, thickness (typically one to several tens of micrometers), microhardness, and adherence of layers are heavily dependent on the laser pulse repetition rate, the pulse duration, and the pressure of the ambient gas. The microhardness of films in the center of the spot was typically increased by a factor of 2-3 over the values of bulk Ti and Zr. The best nitrified layers on Zr substrates reached approximately the hardness of ZrN.

6.1.2 Semiconductors

Surface oxidation and nitridation of semiconductors has been investigated mainly for Si and, to a lesser extent, for GaAs and InP, see Table 6.1. Comprehensive reviews have been given by BOYD ⌊6.63,119⌋, TOKUYAMA et al. ⌊6.64⌋ and SIEJKA et al. ⌊6.18⌋.

a) Silicon

Surface *oxidation* of c-Si and a-Si has been investigated for cw and pulsed laser irradiation in air and O_2. Oxide formation by annealing of O^+ implanted Si surfaces has also been demonstrated. In most of the experiments, the Si

Table 6.1. Laser-induced/enhanced surface modification (mainly oxidation and nitridation) of inorganic materials. The surface composition of the processed material refers to the main component formed; layers where this composition was not analyzed in detail are denoted by ox. The abbreviations employed are listed at the end of the book (see also caption to Table 5.1). The given growth rates, W, denote average values. h_L is the thickness of the original free standing or evaporated film, while h is the thickness of the oxide layer formed by cw or pulsed irradiation. Here, thin native oxide layers that spontaneously form on various substrate materials were not taken into account

Surface composition	Substrate h_L [μm]	Parent/Carrier [10^3 mbar]	Laser λ [nm; μm]	W [μm/s] (I [W/cm²]); ϕ [J/cm²]) h [μm]	Remarks	Ref.
CdO	0.1–0.2 Cd/glass[a]	1 air	Ar, ML Kr	$h \leq 0.2$ (100)	P, VIS	[6.3–6]
Cr₂O₃	0.1–0.3 Cr/SiO₂	1 air	1054 p-Nd:glass	$h = 0.012$ (1E−3)	ρ, ESCA; at high intensities 5% CrO₃	[6.7–9]
Cr₂O₃	0.02–0.06 Cr	1 air	10.6 CO₂		P, TEM	[6.10]
Cu₂O	0.1–0.2 Cu/glass	1 air	Ar, ML Kr	$h \leq 0.2$ (100)	VIS	[6.3–6]
Cu₂O, CuO	300 Cu	1 air	10.6 CO₂	$h = 0.05$ (100)	SEM, TEM	[6.11–15]
Cu₂S	Cu	0.1 CS₂	515 Ar			[6.16]
Ga₂O₃₋δ + As₂O₃₋ε	GaAs	O₂	ArF			[6.17–19]
Ga₂O₃ + As₂O₅	GaAs	O₂	185, 254 UV lamp			[6.17, 18]
GaAs ox.	GaAs		488 Ar		HS 280–600, AES	[6.20]
GaAs ox.	(110)n-, p-GaAs	10^{-8}–10^{-3} O₂	515 Ar		p, photoemission	[6.21]
GaAs ox.	(110) GaAs	1 O₂	515 Ar		p, photoemission	[6.22]
Ga₂O₃ + As₂O₃	GaAs, a-GaAs	0–5 O₂	p-532 Nd:YAG Ruby		P, p, RBS	[6.23, 24]
α-Ga₂O₃ + As₂O₃	(100) GaAs	1–5 O₂	Ruby		ε_d, AES, RED	[6.25]
InAs ox.	InAs	488 Ar			HS 280–600, AES	[6.20]
InGaAsP ox.	InGaAsP					
In₂O₃ + InPO₄	(100)n-InP	10^{-2} N₂O	ArF	$W = 5E-5$ $h = 0.015$	⊥, ∥, HS 620–690, P, p, SEM, XPS, X	[6.26, 27]
InP ox.	InP	488 Ar			HS 280–600, AES	[6.20]
Nb₂O₅₋δ	130 Nb	1 O₂	10.6 p-CO₂ XeCl	$W = 40$ Å/p (0.79)	P, XPS, ESCA, LPPC, XPS; many suboxides	[6.28, 29]
δ-NbN₁₋ε (β-Nb₂N₁₋δ)	Nb	Liquid N₂	Ruby	$h = 1$	ρ, AES, X	[6.30]

Table 6.1 (continued)

Surface composition	Substrate h_L [µm]	Parent/Carrier [10^3 mbar]	Laser λ [nm; µm]	W [µm/s] (I [W/cm²]; ϕ [J/cm²]) h [µm]	Remarks	Ref.
SiO₂	SiO/Si SiO/C	1 air	ArF	$h \leq 0.4$	P, IR	[6.31]
SiO₂	Si	10^{-2} N₂O	ArF	$W = 5E-6$	∥	[6.26, 27]
SiOₓ	Si	1 air	KrF		Dependence of x on dopant	[6.32]
SiO₂	(100), (111) p-Si	1 O₂, 1 air	266 Nd:YAG	$h = 0.03$	IR, TEM, AES	[6.33]
SiO₂	(100) Si	1 O₂	XeCl	$W = 0.01$ $h = 1$	HS 670, n, PP, IR, XPS; MOS device	[6.34, 35]
SiO₂	(100) Si	≤0.7 O₂ + 0.015 Cl₂	XeCl	$W = 0.1$ Å/p $h = 0.35$	HS 600, p, SEM, AES	[6.47]
SiO₂	(100), (111) Si	1 O₂	458, 488, 515 Ar		HS 1170, EM 3–30, EM(100)>EM(111), P, λ	[6.36]
SiO₂	(100) n-Si	1 O₂; 0.3 O₂ + 0.7 H₂O	ML Ar, 351–753 Kr	$W \leq 2E-5$	HS 1050–1170, EM 40–60, λ, n	[6.37, 38]
SiO₂	Si	1 O₂	ML Ar	$W = 5E-4$	HS 670, P, T	[6.39]
SiO₂	Si	1 O₂	515 Ar			[6.40]
SiO₂	(111) Si	1 O₂	Ruby	$h \leq 7E-4$	Al, P, p, AES, LEED, EELS	[6.41]
SiO₂	(100) Si	10^{-5}–8 O₂ 8 CO₂	Ruby	$W = 1E5$ (3.5) $h = 0.05$	P, p, K, RBS; surface melting	[6.42, 43]
SiOₓ	a-Siᵇ/(100) p-Si	1 air	Ruby		TEM	[6.44]
SiO₂	a-Siᵇ/Si	O⁺ impl.	1054 Nd:glass		IR, TEM, SIMS	[6.45]
SiOₓ	Si, SiO₂/Si	1 Ȯ₂	1064, 532 p-Nd:YAG		SIMS; surface melting	[6.46]

SiO$_2$	Si	1 air	488–514 Ar +CO$_2$	$h = 0.1$	Twin beam	[6.47]
SiO$_2$	(100) Si	O$_2$, H$_2$+O$_2$	9–11 CO$_2$	$h = 0.18$	HS 600–700, IR; 8 MV/cm	[6.48, 49]
SiO$_2$	a-Si^b/Si	1 O$_2$	10.6 cw CO$_2$		IR, RHEED	[6.50]
SiO$_2$	Si	1 O$_2$, 1 air	Ar, 10.6 CO$_2$	$W = 1E-3$	HS 570–620, P, IR	[6.51]
Si$_x$N$_y$	(100) Si	NH$_3$	ArF	$h = 2.5E-3$	EF, HS 670, AES, K	[6.52]
Si$_x$N$_y$	(100) Si	$10^{-3}-8$ N$_2$, NH$_3$	Ruby		P	[6.43]
SnO	0.2 Sn/SiO/glass^a	1 air	Ar		P, TEM	[6.53]
TeO$_2$	Te/glass	1 air	ML Kr	$h = 0.1$ (100)	P	[6.5, 6]
Ti$_x$O$_y$	0.15 Ti/c-Al$_2$O$_3$; 0.15 Ti, Au/c-Al$_2$O$_3$	1 air	p-Nd:YAG	(0.1–2)	P, ρ, K, X	[6.53, 54]
Ti$_2$O$_3$	Ti$_{50}$Zr$_{10}$Be$_{40}$	$10^{-7}-10^{-9}$ O$_2$	531 Kr	(≤ 0.015) $h = 0.1$	RA, SEM, X	[6.57]
TiN$_x$O$_y$	Ti	1 air, N$_2$+1% O$_2$, liquid N$_2$	10.6 cw, p-CO$_2$	$h \leqslant 3$	P, SEM, TEM, X; hardness, nonuniform $0 \leqslant x \leqslant 1$, $0 \leqslant y \leqslant 2$	[6.54–56]
V$_2$O$_5$	0.1 V/glass^a	Air	Ar	$W = 10$ ($\leqslant 3E3$) $h = 0.1$	P, TEM	[6.58]
V$_2$O$_5$	V	1 air	10.6 CO$_2$		K, TEM	[6.59]
ZrN$_x$O$_y$	Zr	1 air, N$_2$+1% O$_2$, liquid N$_2$	10.6 cw, p-CO$_2$	$h = 1-30$	P, SEM, TEM, X; hardness, nonuniform $0 \leqslant x \leqslant 1$, $0 \leqslant y \leqslant 2$	[6.54–56, 60]

^{a} Glass of complex composition or unspecified. Quartz glass, for example, is denoted by SiO$_2$, and crystalline quartz by c-SiO$_2$. ^{b} Amorphization by ion implantation.

substrate was preheated to several hundred degrees (Table 6.1). For sub-band-gap radiation $\lfloor h\nu < E_g(Si) = 1.12$ eV $\hat{=}$ 1107 nm$\rfloor$ the enhancement in the oxidation rate is primarily determined by the laser-induced surface temperature rise. For ultraviolet and visible laser radiation that generates electron-hole pairs, nonthermal contributions to the reaction rate have been observed; these are particularly significant at low laser powers.

Oxidation by CO_2 laser irradiation is based on substrate heating due to free carrier absorption. Preheating of the substrate increases the carrier density and thereby accelerates the heating process (Sect.2.1.1). Films up to about 0.2 μm thick have been grown at an average rate of about $3\cdot10^{-5}$ μm/s. Films with thicknesses > 0.04 μm exhibited an average dielectric breakdown of about $6.5\cdot10^6$ V/cm $\lfloor 6.39,48,49\rfloor$.

The enhancement of the thermal oxidation rate of Si in O_2 by visible and UV laser light irradiation has been studied by a number of groups. SCHAFER and LYON $\lfloor 6.37,38\rfloor$ found that low power visible laser irradiation increases the oxidation rate of Si by approximately 40% while the increase for UV radiation at 350 nm was about 60%, over the temperature range of 1050-1170 K. Although the effect could be of largely thermal origin, it was suggested that at least part of the enhancement is due to electron-hole pair generation at the Si/SiO_2 interface. This suggestion was confirmed by YOUNG and TILLER $\lfloor 6.36\rfloor$, who found the oxidation enhancement to increase linearly with both laser beam intensity and wavelength. The enhancement is greater for (100) than for (111) Si, which is the opposite of what one would expect from purely thermal oxidation. With increasing laser irradiance, quantum effects are washed out and laser-induced substrate heating becomes the dominating oxidation mechanism.

Oxidation of Si by XeCl laser irradiation has been studied by ORLOWSKI et al. $\lfloor 6.34,35\rfloor$. Here, particular emphasis was placed on the investigation of the physical properties of films with regard to semiconductor device applications. The oxide layer thickness was found to increase initially with laser beam illumination time at a rate of 0.01 μm/s (τ = 5 ns, repetition rate 100 Hz). Above a layer thickness of 0.2 μm, the growth rate decreased by about a factor of 3. This behavior is in qualitative agreement with the model mentioned above. It should be emphasized, however, that even in this later phase of growth, oxidation is enhanced by a factor of 30 over standard thermal oxidation at 1300 K in 1000 mbar O_2 (when taking into account that laser heating occurs only during single pulses, the enhancement factor is much higher). A sharp onset to rapid oxide formation was observed at laser pulse energy densities near 0.35 J/cm^2, where surface melting occurs.

Characterization of the oxide films was performed by IR spectroscopy, XPS, and capacitance-voltage (CV) measurements. The latter experiments revealed fixed oxide charge densities of $3 \cdot 10^{11}$-$8 \cdot 10^{11}/cm^2$ and surface state densities of the same magnitude. The leakage currents were, typically, 10^{-6} A at $3 \cdot 10^5$ V/cm. A considerable improvement in the electrical quality of as-grown oxide films was found to be possible by means of a short (20 min) anneal at 1170 K in 1000 mbar O_2. After this anneal the fixed charge density was about $6 \cdot 10^{10}/cm^2$ and the surface state density $2 \cdot 10^{11}/cm^2$ eV. Leakage currents were below the detection limit of 10^{-11} A (at $5 \cdot 10^5$ V/cm). The breakdown voltage was in all cases > $5 \cdot 10^5$ V/cm.

The oxidation of Si in O_2 and CO_2 atmospheres by high intensity ruby laser pulses ($\phi \geqslant 2$ J/cm^2, $\tau \approx 15$ ns) sufficient to induce surface melting, has been studied by BENTINI et al. $\lfloor 6.42,43 \rfloor$.

Another development in Si surface oxidation is twin-beam irradiation, where, for example, an Ar$^+$ laser produces an excess of free carriers in the conduction band such that the CO_2 laser radiation can be absorbed more efficiently, see Sect.2.1.1. This possibility permits improved localization of the heating process.

Surface *nitridation* of Si preheated to 670 K has been performed in an NH_3 atmosphere by ArF laser radiation $\lfloor 6.52 \rfloor$. The nitridation process is characterized by an initial rapid growth (during about 2000 pulses of 12 ns duration with an energy of 15 mJ/cm^2), followed by inhibited growth similar to thermal nitridation. The maximum film thicknesses achieved were $2.5 \cdot 10^{-3}$ μm. On the basis of AES studies, the laser grown films are very similar to those thermally grown at 1300 K. The nitridation process is believed to be related to NH_2 radicals produced by photodissociation of NH_3 (Sect.2.2.1). The NH_2 radicals easily react with the Si surface and nitridation may proceed similarly to the oxidation process.

b) Compound Semiconductors

The laser-enhanced oxidation of III-V compounds, and in particular of GaAs, with visible and UV laser light has met with increasing interest (see Table 6.1).

Low intensity 514.5 nm Ar$^+$ laser irradiation ($\leqslant 3$ W/cm^2) of GaAs exposed to O_2 increases the sticking probability for oxygen by a factor of 10^3 and leads to the formation of an oxide layer consisting of Ga_2O_3 + As_2O_3. Similar experiments have been performed also with low intensity ArF laser radiation ($\phi = 0.045$ J/cm^2, $\tau = 15$ ns) and UV lamp radiation (2.5 mW/cm^2 at 185 nm and 15 mW/cm^2 at 254 nm) $\lfloor 6.17,18 \rfloor$. While the stoichiometry of the lamp-formed oxide was $Ga_2O_3 + As_2O_5$, the laser-formed oxide showed an oxygen deficiency in comparison with the overall formula $Ga_2O_3 + As_2O_3$. This behavior is not well understood.

Oxidation of various III-V compounds under high intensity visible laser light irradiation has been studied by COHEN et al. [6.23] and by FUKUDA and TAKAHEI [6.20].

Photo-enhanced oxidation of Si and InP in a N_2O atmosphere excited by ArF laser light at parallel and combined perpendicular/parallel incidence has been demonstrated by ZARNANI et al. [6.26,27]. For InP preheated to 685 K, the oxidation rate at parallel incidence was $5 \cdot 10^{-5}$ µm/s (10 mbar N_2O). The growth rate increased with increasing N_2O pressure. This shows the importance of excited oxygen, which is formed according to (2.27). Combined perpendicular/parallel irradiation increased the oxide layer thickness by about 20%. This additional enhancement is due to electron-hole pair generation and possibly also due to direct substrate heating. The composition of the oxide layers contained In_2O_3 and a phosphate, probably $InPO_4$.

c) Oxidation Mechanisms

From the existing literature it is very clear that the photo-enhanced oxidation of semiconductors under visible and UV irradiation is not a purely thermally controlled process. Nonthermal contributions are essential, particularly at low laser irradiances. It seems to be clear also that various different nonthermal mechanisms contribute to the oxidation process and that their relative importance depends on the particular system under investigation and the laser wavelength and irradiance. Among these mechanisms are:

- The migration of photogenerated electrons through the oxide layer. This may result in both the formation of oxygen anions and a strong electric field, which in turn pulls the oxygen ions through the oxide layer. The oxidation rate decreases with increasing oxide layer thickness due to the reduction in field strength. Such a field-assisted diffusion process would at least qualitatively explain the dependence of the oxidation rate, which has been investigated mainly for Si. This mechanism also seems to be essential in the Si-halogen etching reactions that are described in Sect. 8.2 [6.65,66].

- For UV and for far UV laser light, the photo-induced formation of oxygen vacancies within the oxygen layer may become important [6.17-19,67,68]. Oxygen-vacancy-related defects increase the mobility and thereby the transport of the oxygen.

- Far UV radiation also increases the formation of active oxygen species (O, ozone). These can be produced either in adlayers or within the ambient medium. Excited oxygen atoms $O(^1D)$ may directly react with semiconductor surfaces to form oxide layers [6.26,27].

Similar mechanisms are expected also for nitridation reactions.

In summary, one can say that surface oxidation and nitridation result in the formation of films with thicknesses of, typically, 0.1-1 μm. Such films can be produced either by single-step direct writing, or over extended areas of up to several square centimeters. The technique is complementary to laser-induced gas-phase deposition (Sect.5.3), which makes it possible to grow uniform films which are, typically, between 10 and 10^3 times as thick.

6.2 Reduction, Metallization

Just as laser-induced heating of certain materials in an oxygen-rich atmosphere encourages the oxygen to become incorporated in the lattice, specific materials can, in contrast, be persuaded to give up their oxygen to a reducing environment under the appropriate laser-heating conditions. This has been demonstrated for oxidic perovskites, e.g. $BaTiO_3$, $PbTiO_3$, $PbTi_{1-x}Zr_xO_3$ (PZT), $Pb_{1-3y/2}La_yTi_{1-x}Zr_xO_3$ (PLZT) and $SrTiO_3$, by BÄUERLE and co-workers [6.69-73]. These materials are insulators with a band gap of, typically, 3 eV. They are ferroelectric and thereby piezoelectric. It is well known that the physical properties of these materials can be dramatically changed by oven heating of the bulk material, e.g. in H_2 atmosphere at 500 - 1500 K [6.74-79]. This treatment results in the generation of oxygen vacancies and free or quasi-free electrons. The concentration of oxygen vacancies and free electrons increases with increasing reduction temperature and with decreasing oxygen partial pressure. The oxygen vacancies act as shallow donor levels and the originally insulating material becomes an n-type semiconductor. The originally transparent material becomes blue to black, depending on the concentration of vacancies [6.74,75]. Because of the fundamental role of the oxygen ion in connection with the dynamical properties of perovskites, the oxygen vacancies also strongly influence the structural phase transitions (ferroelectic and nonferroelectric) observed in these materials [6.76-83].

Laser light irradiation of oxidic perovskites in a reducing atmosphere can result in local reduction of the material surface [6.70-73,84]. While for sub-band-gap radiation ($h\nu < E_g$) the reduction mechanism is mainly thermal, UV and far UV radiation ($h\nu > E_g$) may directly generate oxygen vacancies and quasi-free electrons. The reduction process is reversible, i.e. on heating the material in an O_2 atmosphere or in air, the reduced (blue to black)

Optical micrograph of stripes produced with laser powers P = 75 mW, 125 mW and 190 mW (left to right). The scanning velocity was v_s = 25 μm/s and the hydrogen pressure $p(H_2)$ = 500 mbar. The distance between centers of stripes is 200 μm (after ⌊6.72⌋)

regions vanish. Only a small change in surface morphology is observed. With increasing laser irradiance the degree of reduction increases and the electrical properties of laser-treated regions change from semiconducting to metallic. When the laser intensity is even further increased, beyond a certain value, etching or cutting of the material is observed (Sect.8.3).

Laser-induced reduction and metallization of oxidic perovskites allows single-step conductive patterning of the otherwise insulating materials' surfaces. Metallization has been studied in detail for hot-pressed optically transparent ferroelectric PLZT ceramics ⌊with 9.5 at. % La (y = 0.095) and a Zr:Ti ratio of 65:35 (x = 0.35)⌋ by KAPENIEKS et al. ⌊6.72,73⌋. The electrical resistivity of the bulk material was > 10^{14} Ωcm.

Figure 6.2 shows an optical micrograph of metallic stripes produced with increasing power (from left to right) of the UV multiple line output of a Kr^+ laser (λ = 337-356 nm) on the surface of a transparent PLZT sample. After laser treatment the sample was rinsed in dilute HNO_3 in order to remove the Pb condensed near the edges of the stripes.

In Fig.6.3, the resistivity per unit length of stripes is plotted versus laser power. The different symbols refer to a freshly prepared sample and a sample annealed at about 1100 K. Initially, both samples show an approximately exponential decrease in resistivity with increasing laser power, and then a decrease in slope in the semilogarithmic plot. For laser powers above about 180 mW, microcracks are occasionally observed in the surface region adjacent to the metal stripe. These cracks have no influence on the electrical conductivity of the stripes. However, when the laser power is increased to above 250-300 mW, cracks that penetrate deeply into the bulk PLZT material underneath the metallic stripe are observed. Therefore, this regime was not investigated in any more detail. The single data points shown in the figure for the same laser power were obtained from resistivity measurements on the same stripe, but for different locations of electrodes. The scattering of data is probably due to inhomogeneities in the transformed material, which may possibly be related to inhomogeneities in the bulk material. The results presented in the following were all obtained with freshly prepared samples.

146

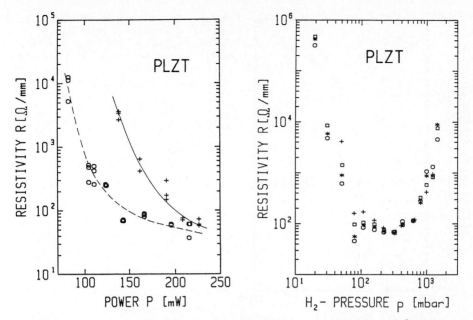

Fig.6.3. Resistivity of laser-processed stripes as a function of laser power with v_s = 25 μm/s and $p(H_2)$ = 500 mbar. o: PLZT 9.5/65/35 freshly prepared; +: Sample annealed at 1100 K (after ⌊6.72⌋)

Fig.6.4. Resistivity of laser-processed stripes as a function of H_2 pressure. P = 200 mW, v_s = 25 μm/s. o: Resistivity measured immediately after laser processing and Cu electrode evaporation; ⋆: Resistivity after 20 h; □: After 50 h; +: After annealing the sample at 373 K for 1 h (after ⌊6.72⌋)

Figure 6.4 shows the resistivity of stripes as a function of H_2 pressure for different time intervals of resistivity measurements. Here, each data point represents an average of five resistivity measurements. A pronounced minimum in resistivity for pressures between about 100 and 600 mbar can be observed. It can be seen that ageing influences the resistivity of stripes produced at lower H_2 pressures more strongly than those produced at higher pressures. The dependence of the resistivity of stripes on the scanning velocity is shown in Fig.6.5. Again, a sharp minimum is observed for a certain range of scanning velocities, at otherwise constant parameters. The change in resistivity is about 5 orders of magnitude on both sides. For scanning speeds below about 10 μm/s, scanning electron micrographs show many cracks in the stripes. This explains the broad scattering of data points in this region. At higher scanning velocities, no cracks can be observed and the reproducibility of measurements is within about a factor of two. This is quite satisfactory in view of the wide range of resistivities that can be obtained for different processing parameters.

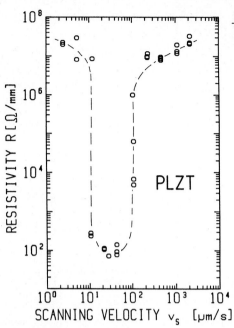

Fig.6.5. Resistivity of stripes as a function of scanning velocity. P = 190 mW, p(H$_2$) = 500 mbar (after ⌊6.72⌋)

PLZT

RESISTIVITY R [Ω/mm]

SCANNING VELOCITY v_s [μm/s]

From the foregoing results it has become evident that ranges of optimal processing parameters exist, for which the resistances of stripes are typically of the order of 100 Ω/mm. In these ranges of processing parameters, the stripes have metallic conductivity. This can be concluded from a simple estimate of the specific resistivity of the stripes, which yields values of about 10^{-4} Ω cm.

The conductivity of metallized regions within PLZT surfaces produced by UV laser irradiation in H$_2$ atmosphere is essentially determined by: the reduction of the material to metallic Pb, Ti and Zr, the evaporation of Pb and, for certain parameters, the cracking of stripes. The evaporation of Pb results in the occurrence of a shallow groove in the middle of the metallic stripe. This local depletion of Pb is consistent with X-ray microanalysis and similar investigations performed on PbTi$_{1-x}$Zr$_x$O$_3$ ⌊6.71⌋. The dependencies of the resistivity shown in the figures can be qualitatively understood in the light of the aforementioned points: An adequate processing time (~ $1/v_s$) and a sufficient amount of H$_2$ are necessary for the metallization of the PLZT surface. On the other hand, if v_s is too small, an increase in Pb evaporation and a subsequent cracking of stripes is observed (Fig.6.5). For high H$_2$ pressures the mean free path of the evaporating Pb is diminished and the dense Pb vapor attenuates the laser beam intensity. This causes the increase in resistance (Fig.6.4).

The increase in power threshold observed for annealed samples (curve indicated by + symbols in Fig.6.3) can be understood from the shift in band edge caused by the annealing. This shift in band edge is a well-known phenomenon in PLZT. In fact, one can see that the characteristic yellowish color of freshly prepared PLZT is faded out in annealed samples. Because the energy of the laser light is only near the band-gap energy of the PLZT, a small shift in the band edge to higher energies will cause a significant change in absorbed laser power.

Large-area electrodes (0.5×0.5 cm^2) have been prepared in a similar way on PLZT surfaces. These electrodes were characterized by temperature-dependent dielectric measurements performed on samples with different thicknesses (0.1-2 mm) at frequencies from 10 kHz to 10 MHz ⌊6.73,120⌋. Below 400 K the laser-fabricated contacts led to higher dielectric constants than conventional evaporated Au electrodes. The difference is most pronounced for small sample thicknesses. Thus, in combination with the observed increase in adherence, such laser-processed electrodes could be superior to conventional electrodes in miniaturization problems.

6.3 Doping

This section deals with the doping of semiconductor surfaces. It does not, however, contain any material on the deposition of doped semiconducting films, as this has already been discussed (Chap.5). Laser-induced doping of materials' surfaces takes advantage of the high heating and cooling rates that can be achieved with pulsed lasers or scanned cw lasers (Sect.2.1.1). The short temperature cycles obtained with lasers make it possible to produce very shallow heavily doped layers within solid surfaces. While pulsed lasers are mainly used for large-area (sheet) doping, cw lasers allow direct writing of patterns with lateral dimensions down to the submicrometer level (Sect.4.2). Table 6.2 lists the systems investigated so far.

Laser doping is, in general, performed at normal incidence. The absorbed fluence must be high enough to substantially heat or even melt a thin layer of the substrate surface in order to allow dopant incorporation by high temperature or liquid-phase diffusion. The source of dopant atoms may be a thin layer of an element implanted into or evaporated onto the substrate surface, or adsorbed-, gas-, liquid-, or solid-phase parent molecules, which are photothermally or photochemically decomposed by the laser light. Therefore, laser doping often involves two simultaneous steps: heating or melting of the substrate surface and photodecomposition of the parent

Table 6.2. Laser-induced doping of semiconductors. The abbreviations employed are listed at the end of the book (see also caption to Table 5.1). The table does not contain the deposition of doped films, which is included in Table 5.1. The following additional contractions have been used: impl. for ion implantation of species with subsequent laser treatment; evap. for doping from thin evaporated film; n_c for dopant concentration; and h_j for junction depth

Solid	Parent/Carrier	Laser	n_c [cm^{-3}] h_j [μm]	(I [W/cm^2]; ϕ [J/cm^2])	Remarks	Ref.
GaAs:S	H$_2$S	ArF, XeF	10^{19} 0.15	(0.1) (0.3)	P, λ, PP, SIMS; 10.8% solar cell	[6.85 – 87]
GaAs:Se	H$_2$Se/H$_2$+AsH$_3$	p-532, 1064 Nd:YAG	10^{21} 0.02	(0.12)	HS 300–900, P, p, ρ, SIMS; ohmic contact	[6.88 – 90]
GaAs:Si	SiH$_4$	p-532 Nd:YAG	$>10^{19}$ 0.5		SIMS	[6.23]
GaAs:Zn	Zn(C$_2$H$_5$)$_2$/H$_2$ + AsH$_3$	p-532, 1064 Nd:YAG	$3 \cdot 10^{20}$	(0.15)	HS 300–900, P, p, AES, RBS, SIMS; ohmic contact	[6.88, 90]
InP:Cd	Cd(CH$_3$)$_2$	ArF		(2.7)	S, ρ, AES; ohmic contact	[6.91]
InP:Cd	Cd(CH$_3$)$_2$	257 SH + 515 Ar	$>10^{19}$	$2 \cdot 10^2 + 2 \cdot 10^6$	L, P, λ, p, ρ; ohmic contact, pn junction	[6.92]
InP:Zn	Zn(CH$_3$)$_2$	ArF			S, ρ, AES; ohmic contact	[6.91]
Si:Al	Al$_2$(CH$_3$)$_6$	ArF	10^{20} 0.12	(0.12)	ρ, SIMS; solar cell	[6.93]
Si:Al	Al evap.	Ruby	$2.7 \cdot 10^{20}$ <0.4	(1.5)	P, ρ, RBS; solar cell	[6.94]
Si:Al	Al evap.	p-Nd:YAG		$3 \cdot 10^{-3}$	ρ; pn junction	[6.95]
Si:B	BCl$_3$/H$_2$	257 SH, 515 Ar	10^{21} 0.3 – 1	(0.1)	L, ρ, SEM, SIMS	[6.96]
Si:B	BCl$_3$	ArF, XeF	$3 \cdot 10^{20}$ 0.35	(0.7)	AL, P, p, ρ	[6.86, 87, 97]
Si:B	B(CH$_3$)$_3$	ArF	$1.7 \cdot 10^{21}$ 0.7	(0.23)	ρ, SIMS; 9.6% solar cell	[6.93]
Si:B	B(CH$_3$)$_3$	ArF	$4 \cdot 10^{19}$ 0.4	(1)	ρ, SIMS	[6.98]

Material	Dopant source	Laser	Conc. (cm⁻³) / (E) / depth	Notes	Ref.
Si: B	B(C$_2$H$_5$)$_3$	ArF, XeF	$>10^{20}$	ρ, ρ, AES	[6.99]
Si: B	B$_2$H$_6$	XeCl	0.05 – 0.8 (1)	AL, P, ρ, SIMS	[6.125]
Si: B	Emulsitone XB-100[a]	ArF, KrF	$5 \cdot 10^{20}$ 0.08	Spun-on film	[6.97]
Si: B	B evap.	Ruby	$5 \cdot 10^{21}$ (1.5) 0.4	ρ, TEM, SIMS; pn junction	[6.100, 101]
Si: Bi	Bi evap.	Ruby	$4.5 \cdot 10^{19}$ (1.5) 0.3	ρ, RBS; solar cell	[6.94]
Si: Ga	Ga evap.	Ruby	$2.1 \cdot 10^{20}$ (1.5) 0.3	ρ, RBS; solar cell	[6.94]
Si: In	In evap.	Ruby	$2.4 \cdot 10^{19}$ (1.5) 0.3	ρ, RBS; solar cell	[6.94]
Si: P	PCl$_3$	ArF	10^{20} (1) 0.3	ρ, SIMS; npn structures	[6.98]
Si: P	PCl$_3$	ArF	(0.8)	ρ	[6.86, 87]
Si: P	PH$_3$/Ar	Alexandrite	10^{20} (2.5) 0.14	ρ, SIMS; solar cell	[6.102]
Si: P	C$_{12}$H$_{27}$O$_4$P/sol.	Ruby		ρ, SIMS, RBS; 13% solar cell	[6.103]
Si: P	P evap.	Ruby	$4.5 \cdot 10^{19}$ (1.5)	solar cell	[6.94]
Si: P	PF$_5$ impl.	p-Nd: YAG		ρ, SEM, EBIC; 11% solar cell	[6.104]
Si: Sb	SbCl$_3$/C$_2$H$_5$OH sol.	Ruby	$4 \cdot 10^{20}$ (1.5) 0.2	ρ, SIMS, RBS	[6.103]
Si: Sb	Sb evap., impl.	KrF	$2 \cdot 10^{21}$ (1) 0.35	PP, TEM, RBS	[6.105]
Si: Sb	Sb evap., impl.	Ruby	10^{21} (2)	ρ, RBS; solar cell	[6.94]
a, p-Si: B	BCl$_3$	ArF, XeF	<0.3	AL, P, λ, ρ; a-, p-Si/2000 Å SiO$_2$/Si	[6.86, 87]

[a] Emulsitone and XB-100 are supplied by Allied Chemical, USA.

molecules. Clearly, mass transport may be determined either by the supply of
doping atoms at the substrate surface or by the diffusion of the dopant
within the surface of the material. When the sample surface is melted under
the laser irradiation, the transport of the dopant to the surface is usually
the rate-limiting step because of the extremely rapid diffusion of atoms in
the liquid phase. For special substrate-dopant combinations this may also
hold for solid-state diffusion well below the melting point. Examples for the
latter case are polycrystalline materials and III-V semiconductors, where
specific dopants such as Cd, S, Se, and Zn can rapidly diffuse into the
material's surface even at moderate temperatures. In such systems the dopant
density within the solid surface can be controlled via the rate of arrival of
free dopant atoms at the substrate surface. This can be realized, for
example, by using inert buffer gases or dilute solutions, which make it
possible to reduce the molecular diffusion rates to the surface. The
diffusion depth, on the other hand, can be controlled independently by
adjusting the surface temperature and the time of heating. It can be seen
from Table 6.2 that most of the doping experiments have been performed for
Si. The reason is based not only on the special interest in this material,
but also on Si's great stability, which allows a much wider range of laser
operating parameters than, for example, for III-V semiconductors. For the
latter, decomposition at the surface can occur by selective evaporation of
the more volatile component. Additionally, laser-induced heating of III-V
semiconductors can introduce slip planes and other defects which degrade the
electrical properties of these materials.

6.3.1 Sheet Doping

The various possible dopant sources mentioned above have all been proved for
large-area doping of Si. Sheet doping of compound semiconductors has been
demonstrated mainly with gas-phase precursor molecules.

a) Silicon

Doping of Si from *adlayers* has been demonstrated by DEUTSCH et al.
⌊6.86,87,97⌋. The Si sample was first exposed to a 133-mbar atmosphere of
BCl_3, which was pumped off before irradiation with XeF laser light. The sheet
resistance achieved was about 650 $\Omega/\square$. Similar experiments were performed
with PCl_3. Because adlayers provide a strictly finite source of dopant,
extremely flat doping profiles can be produced by this technique. Clearly, a
prior condition for the application of this technique is a strong enough bond
between the parent molecules and the substrate surface, so that adlayers are

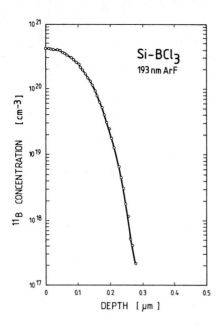

Fig.6.6. SIMS profile of the B concentration within a (100) Si surface. ArF laser radiation and 67 mbar BCl_3 were used for doping (after [6.86])

formed that do not immediately desorb after gas-phase exposure. This is, of course, not the case for many adsorbate-adsorbent systems. For example, doping experiments using H_2S and GaAs substrates were unsuccessful [6.86,87].

Gas-phase doping of Si has been demonstrated for a number of different dopant gases. Detailed experiments have been performed in a BCl_3 atmosphere with XeF and ArF lasers. Because 351 nm XeF radiation is not absorbed by BCl_3 molecules, pyrolytic decomposition of the parent molecules at the gas-solid interface is the most likely mechanism. On the other hand, ArF laser radiation photodissociates BCl_3. Figure 6.6 shows a SIMS profile of the B concentration for a (100) n-type Si wafer. The fluence of the ArF laser radiation was 0.6 J/cm^2 and the BCl_3 pressure 67 mbar. The scanning velocity of the line focus used in this experiment (0.3 mm x 10 mm; see also Fig.4.2b) was about 100 μm/s. The laser pulse repetition rate was 10 pps. The junction depth estimated from the figure is about 0.35 μm. The B concentration at the Si surface is in excess of the solid solubility of B in Si. Similar experiments were performed for polycrystalline Si and for 5000 Å amorphous CVD Si on 2000 Å SiO_2/Si substrates (it should be noted that under the fluences involved, recrystallization of the initially amorphous Si into fine grain polycrystalline Si will occur simultaneously with doping) [6.86,87]. Figure 6.7 shows the sheet resistances of c-Si and a-Si as a function of ArF and XeF laser fluences. The lowest sheet resistance obtained with ArF laser

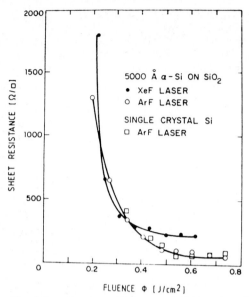

FLUENCE Φ [J/cm²]

Fig.6.7. Sheet resistance of B-doped c-Si and a-Si as a function of ArF and XeF laser fluences. The BCl_3 pressure was 67 mbar (after [6.86])

radiation, 60 Ω/□, is about 1/3 of that obtained with the XeF laser. This result strongly indicates that the photolytic decomposition of parent molecules enhances the doping substantially. The steep rise of the curves at fluences below ~ 0.2 J/cm² is probably due to the threshold for melting of the Si film. From the electrical measurements in the high fluence region and the doping densities and the doping depths, one can estimate that essentially all B atoms introduced by the UV doping process are electrically activated. Due to the high doping density, however, the mobility of carriers in the c-Si sample is strongly reduced (at 300 K about 41 cm²/Vs). Investigations on the dependence of the sheet resistance, and on the thickness of the BCl_3 adlayer on the BCl_3 gas pressure, have revealed the important role of physisorbed molecules in the doping process. The pressure dependence of the sheet resistance was found to be in qualitative agreement with adsorption isotherms, i.e. the sheet resistance decreases up to 70-130 mbar, which is the same pressure region as that in which the thickness of the adlayer saturates.

Liquid-phase doping of Si with P and Sb has been demonstrated by STUCK et al. [6.103]. The doping solutions were either tributyl-phosphate ($C_{12}H_{27}O_4P$) with $2.5 \cdot 10^{21}$ P atoms/cm³, or antimony trichloride ($SbCl_3$) in ethanol with $1.5 \cdot 10^{21}$ Sb atoms/cm³. Irradiation was performed with Q-switched ruby laser light ($\tau \approx$ 20 ns, and $\approx$ 1-2 J/cm^2), which is either only very slightly absorbed by these liquids, or else not at all. The sheet resistances of P and Sb doped layers were 120 $\Omega/\square$ and 60-80 $\Omega/\square$, respectively.

Boron doping of Si was also performed by heating and ablating spun-on *dopant films* (Emulsitone and XB-100 $\lfloor$6.121$\rfloor$) with ArF and KrF laser radiation. The formation of pn-junctions by ruby laser irradiation of thin films of B evaporated on P-doped Si wafers has been described by NARAYAN et al. $\lfloor$6.100,101$\rfloor$. Carrier concentrations of $2 \cdot 10^{15}$ - $3 \cdot 10^{16}$/cm^2 were obtained. Sheet resistivities ranged from 80 to 8 $\Omega/\square$. The corresponding carrier mobilities were 40-24 cm^2/Vs. Thin evaporated films have also been used to dope Si with P, Sb, Bi, Al, Ga and In $\lfloor$6.94,101,105$\rfloor$. The experimental results are in reasonable agreement with model calculations $\lfloor$6.106$\rfloor$ and similar to those obtained by laser annealing of ion-implanted Si surfaces $\lfloor$6.105$\rfloor$. This latter technique has been studied in great detail $\lfloor$6.108-118$\rfloor$ and the set of examples incorporated in Table 6.2 is by no means complete.

Irrespective of whether Si doping is performed from a thin evaporated film or via pyrolytic or photolytic decomposition of adsorbed-, gas- or liquid-phase parent molecules, or from spun-on layers, pulsed laser irradiation seems to require melting of the Si surface in order to allow the dopant to be incorporated by liquid-phase diffusion. This becomes evident when estimating the diffusion length of dopants during typical heating times. Even at temperatures near the melting point of solid Si, the diffusion length of dopants such as B or Al would be only a few Ångströms (see Sects.2.1.1 and 3.4). With stationary or scanned cw lasers, however, the longer dwell times allow solid-phase diffusion that may be enhanced by thermal, stress and dopant concentration gradients $\lfloor$6.96$\rfloor$.

b) *Compound Semiconductors*

Laser-induced sheet doping has also been demonstrated for GaAs and InP $\lfloor$6.85,88-91$\rfloor$. BENEKING et al. have used Q-switched frequency-doubled Nd:YAG laser light (with and without blocking the fundamental line) to dope GaAs with Se $\lfloor$6.88-90$\rfloor$ and Zn $\lfloor$6.88,90$\rfloor$. The Se doping source was gaseous H_2Se diluted in H_2 + AsH_3. Admixture of AsH_3 decreased the surface decomposition. During the 3 ns laser pulse used in these experiments, the surface was heated to temperatures near the melting point. The H_2Se molecules adsorbed on the GaAs wafer thermally decompose and diffuse into the surface. The loss of As during heating is compensated in part by the As supply from adsorbed AsH_3. With pure 532 nm radiation, surface layers less than 0.02 μm thick with over 10^{20} Se atoms/cm^3 have been produced. Thicker dopant profiles can be obtained by additional illumination with the 1064 nm fundamental line. Similar experiments have been performed with $Zn(C_2H_5)_2$ ⌊6.88,90⌋.

6.3.2 Local Doping

Local doping of semiconductors by cw laser direct writing has been demonstrated for lateral dimensions down to submicrometer levels. Boron doping patterns in (111) and (100) Si have been produced by local pyrolysis and/or photolysis of BCl_3 (133-266 mbar) with 515 nm Ar^+ and 257 nm frequency-doubled Ar^+ laser radiation. For 515 nm Ar^+ laser radiation at power levels 10% below melting (~ 0.6 W) a doped linewidth of 0.6 µm was obtained. This width is considerably smaller than the laser focus of $2w_0 \approx$ 2.5 µm. The increase in spatial resolution is consistent with the exponential dependence of the diffusion coefficient on temperature (see Sect.3.5 and Figs.3.6 and 3.7; the activation energy for isothermal diffusion of B in Si is $\Delta E = 3.69$ eV $\hat{=}$ 85.4 kcal/mole).

Direct cw laser doping of InP with Cd and Zn was demonstrated by EHRLICH et al. ⌊6.92⌋. In these experiments 257 nm frequency-doubled Ar^+ laser radiation was used collinearly with the 515 nm fundamental output. By controlling the individual beam intensities it was possible to vary independently the flux of Cd atoms ⌊produced by photodissociation of $Cd(CH_3)_2$ according to (2.20)⌋ at or near the gas-InP interface and the surface heating. In fact, electrical measurements have revealed a linear increase in dopant concentration (up to > 10^{19} Cd atoms/cm^3) with UV flux. The 515 nm radiation, on the other hand, controls the surface temperature and thereby the spatial extent of the doped region. The dominant mechanism for dopant incorporation seems to be solid-phase diffusion. This is in contrast to pulsed laser doping from evaporated or photochemically deposited surface layers (see above) where solid-phase diffusion is not sufficiently rapid to create the required mass transport and where melting of the metal film and the substrate surface seems to be necessary.

7. Alloying, Compound Formation

In this chapter the main emphasis is placed on laser-induced alloying and compound formation starting from solid solutions or from alternating multiple layers consisting of appropriate proportions of the elements. The chapter does not include the deposition of compound materials and of alloys from the gas and liquid phases, which has already been outlined in Chap.5. Also, oxide formation and surface doping were discussed in Chap.6.

Laser-induced solid-phase processing for materials alloying and synthesis can be classified according to two main regimes, which are determined by the irradiation conditions: Pulsed laser processing requires, in general, materials melting and liquid-phase interdiffusion of the constituents. In cw laser processing, the much longer dwell times usually employed allow for solid-phase diffusion and transformation, in many cases.

In the following sections we will first briefly mention the laser-induced formation of metastable materials and then summarize laser-induced chemical synthesis of stoichiometric compounds such as silicides and compound semiconductors. In this chapter in particular, a clear separation of chemical and nonchemical processing has proved extremely difficult.

7.1 Metastable Materials

The formation of metastable materials with lasers has already been extensively reviewed [7.1-3]. The technique is mainly based on the high heating and cooling rates achieved by laser light irradiation. For scanned cw laser heated surfaces the cooling rates are comparable to those achieved in standard splat cooling or melt spinning techniques (about 10^6 K/s). Pulses from Q-switched and mode-locked lasers, on the other hand, allow rates of 10^{10} to 10^{12} K/s to be achieved. These latter rates make it possible to produce glassy alloys which have never before been available. Detailed investigations have been performed for binary alloys consisting of two transition metals, and for combinations of metals with group IV elements.

Among the binary transition metal systems, glasses of Au-Ti, Cu-Ti, Co-Ti, Cr-Ti, V-Ti and Ag-Cu have been produced. Glass formation seems to fail if single-phase crystallization is possible from the melt, or if the glass is unstable at ambient temperatures ⌊7.2⌋. The most extensive investigations on binary alloys of metal-group IV elements have been performed for silicides. Silicides can be produced as glasses but also as polycrystalline or even as epitaxial films. Because of the wide variety of potential applications (Chap.9), the formation of silicides will be discussed in the following section in greater detail.

7.2 Silicides

The formation of silicides with energy beams, such as laser, electron or ion beams, has generated considerable interest. Most of the experiments start out with metal films deposited on Si substrates. In this section we shall briefly summarize silicide formation by scanned cw ⌊7.4-7⌋ and pulsed ⌊7.8-16⌋ laser irradiation. The main analytical techniques that have been employed for product characterization are scanning electron microscopy, transmission electron microscopy, X-ray diffraction, low-energy electron diffraction and Rutherford backscattering.

Silicide formation using a scanning cw laser beam has been demonstrated for Co ⌊7.7⌋, Nb ⌊7.5⌋, Pd ⌊7.4,5⌋ and Pt ⌊7.5⌋. Uniform layers of MeSi, Me_2Si or $MeSi_2$ (Me = metal) consisting of essentially one single phase, have been produced by varying the laser beam intensity ⌊7.4,5⌋. The dominating microscopic mechanism for silicide formation seems to be solid-phase diffusion.

Silicide formation by $pulsed$ laser irradiation has been investigated for: Au ⌊7.1,13⌋, Co ⌊7.16⌋, Cr ⌊7.17,39⌋, Mo ⌊7.16,39⌋, Ni ⌊7.6,9-12,15⌋, Pd ⌊7.6,14,16,39⌋, Pt ⌊7.6,8,39⌋, Ti ⌊7.15⌋ and W ⌊7.18⌋. Excimer, ruby, Nd:YAG and Nd:glass lasers were mainly employed in these experiments. Here, silicide formation occurs, in general, via surface melting and liquid-phase interdiffusion of the metal and the Si. During the rapid solidification, amorphous films, or films with various different Me_xSi_y precipitates surrounded by Si are formed. However, suitable variation of the laser fluence and/or the pulse length and/or the thickness of the metal film, makes it possible to produce uniform layers with a predetermined single phase or average composition of x:y ⌊7.6,11-13⌋. An example for which both multiphase and single-phase silicide formation have been demonstrated is the Ni-Si system. With relatively thick metal layers, typically 0.1 μm, evaporated on the Si wafer, the quantity of material liquefied by the incident laser light is so large that nucleation and growth of various different phases occurs during solidification. Here, simultaneous formation of Ni_2Si, $NiSi$, $NiSi_2$ and $NiSi_3$ was observed ⌊7.15⌋. On the other hand, GRIMALDI et al. [7.11] showed that with thin evaporated metal layers, typically < 0.02 μm thick, and with certain processing conditions, single-phase $NiSi_2$ can even be grown epitaxially onto (100) and (111) Si wafers. The physical properties of these films are very similar to those produced by standard furnace annealing ⌊7.19,20⌋.

There is some evidence to suggest that for low laser fluences, silicide formation may become possible via a quasi-liquid layer formed at the metal-silicon interface when the laser-induced temperature T reaches the lowest eutectic temperature T_e of the binary metal-silicon system. The thickness of the quasi-liquid layer increases as long as $T > T_e$. The growth rate of the reacted layer is of the order of 1 m/s, which is typical for liquid-phase processing. Note that the temperature T can be smaller than the melting temperatures of both the metal and the Si itself. This mechanism applies well for noble-metal and refractory-metal silicides under both laser and electron beam irradiation ⌊7.8,17,39⌋.

7.3 Compound Semiconductors

Laser-induced synthesis of thin stoichiometric crystalline films of compound semiconductors has been pioneered by LAUDE et al. ⌊7.21,22⌋. The films produced include binary compounds such as III-V (AlSb, AlAs) ⌊7.22-28⌋, II-VI (CdTe, CdSe, ZnSe) ⌊7.24,28-31⌋ and IV-VI (GeSe$_2$, GeSe) ⌊7.21,32,33⌋ semiconductors, and even the ternary compound $CuInSe_2$ [7.34-36]. Synthesis has been achieved by irradiating multilayered films consisting of the alternately deposited elements. The proportions of atoms of different elements were adjusted to correspond to the overall stoichiometric amounts of the synthesized film. In order to avoid oxidation or other contamination during processing in air, films were often encapsulated between SiO_x layers. Irradiation was performed mainly with visible cw Ar^+ or Kr^+ lasers and with pulsed dye-lasers.

Synthesis by *cw* laser irradiation is characterized by the occurrence of cells ⌊Ref. 7.21, Fig.2⌋ and the growth of large crystallites. The cells appear as a close-packed earthworm-like structure that is elongated within the surface of the film. The formation of such cells has been studied in some detail for Cd-Se. It was proposed that cell formation requires the melting of

the metal layers and that it is driven by the diminution of the surface
tension at the metal-air interface. In cw laser synthesized films the size of
the crystallites is in the micrometer range. This has been demonstrated by
scanning electron microscopy and X-ray diffraction for free-standing films of
AlSb ⌊7.23-27⌋, $GeSe_2$ ⌊7.21,32⌋ and $CuInSe_2$ ⌊7.34,35⌋. For $GeSe_2$ a
preferential orientation of the c-axis perpendicular to the plane of the film
was observed ⌊7.32⌋. Both supported and free-standing films exhibit good
stoichiometry and optical properties comparable to those of single crystals.
The most detailed studies on cw laser synthesis have been performed for the
Ge-Se system. Here, typical scanning velocities investigated ranged from 1
cm/s to 20 m/s. The dependence of the different microstructures of films
(crystalline and amorphous) on the various experimental parameters was
investigated. Well-defined periodic structures, probably formed by an
explosive chemical reaction, were observed for certain parameter ranges (see
also Sect.5.2.5).

Synthesis by *pulsed* laser irradiation was found to occur above a threshold
energy sufficient to heat the film up to the melting point and to melt a
portion of the elemental constituents. The dependence of the threshold energy
on pulse duration and substrate temperature was investigated for AlSb ⌊7.22⌋.

Another interesting point is the selection of crystal phases via the laser
beam illumination time ⌊7.29⌋. For example, irradiation of Cd-Te and Cd-Se
free-standing films by dye-laser pulses of $\tau = 10^{-6}$ s resulted in a mixture
of hexagonal and cubic CdTe and hexagonal CdSe, respectively. Chopped Ar^+
laser irradiation with pulse durations of $\tau = 3.6 \cdot 10^{-2}$ s, on the other hand,
revealed films of cubic structure only. Thus either type of phase can be
produced by proper control of the laser beam dwell time. In this connection
it is interesting to note that II-VI single crystals are cubic for growth at
low temperatures and hexagonal for growth at high temperatures. The physical
properties of CdTe films produced by pulsed laser synthesis are very similar
to those of single crystals grown by standard techniques. In particular, the
laser-grown material is very pure with electrical impurity concentrations
ranging from only 10^{15} to 10^{17} defects/cm^3. The uncertainty in the
stoichiometry is only about 1%-2%.

Microcrystalline films of semiconducting $Ge_{1-x}Sn_x$ (x ≈ 0.22) have been
synthesized from amorphous RF sputtered films by means of ArF and KrF excimer
laser radiation ⌊7.37⌋. X-ray diffraction, electroreflectance and Raman
spectroscopy have been used for film characterization. These investigations
have confirmed that the films have a crystalline semiconductor band
structure.

160

While the parameters influencing laser-induced compound formation, such as the energy density profiles, laser beam illumination times, heat of reactions, defect formations, surface tensions, films thicknesses, etc., have been investigated in some detail [7.4,11,12,21,22,38], we are far away from an understanding of the microscopic mechanisms involved in laser-induced materials synthesis.

8. Etching, Cutting, Drilling

Many laser-induced chemical reactions and techniques discussed for materials deposition (Chap.5) can be directly applied to materials etching. This was already indicated for the model systems shown in Fig.3.1, where in some cases the course of the reaction can be turned around by simply shifting the chemical equilibrium to the other side or, as in laser-enhanced electrochemical processing, by changing the polarity of the substrate with respect to a counterelectrode. Nevertheless, the reaction pathways for laser-induced deposition and etching may be quite different. In general, chemical etching requires physisorption or chemisorption of reactants, reactions between adsorbed species and atoms or molecules within the solid surface, and desorption of product molecules. Laser light can influence any or all of these processes by photothermal and/or photochemical excitation of gas- or liquid-phase molecules, adsorbate-surface complexes and/or the substrate surface itself. To separate single mechanisms, etching at normal and parallel incidence may be studied. Another type of laser-induced materials removal that is also included in this chapter is laser-induced ablative photodecomposition (APD) of organic polymers. Here, the laser light breaks chemical bonds directly within the surface of the polymer. No excitation of gas-, liquid- or adsorbed-phase molecules is necessary in this case and processing can be performed also in a vacuum. As indicated in the outline of this book given in Chap.1, the present chapter does not include laser machining that is simply based on laser-induced melting and/or evaporation of the material performed within a non-reactive atmosphere.

Table 8.1 summarizes the results available so far on chemical etching of inorganic materials. It becomes evident from the table that most of the work has concentrated on metals and semiconductors. Chemical etching may be performed in the gas phase *(dry-etching)* or in the liquid phase *(wet-etching)*. In dry-etching the precursor molecules most commonly used are halides and halide compounds, while wet-etching is mainly performed in aqueous solutions of HCl, H_2SO_4, HNO_3, H_3PO_4, H_2O_2, KOH, etc., including mixtures of these. High etch rates together with high spatial resolution

Table 8.1. Dry- and wet-etching of inorganic materials. Etch rates, W, correspond to maximum values achieved. Most of the abbreviations employed are listed at the end of the book (see also caption to Table 5.2). The following additional contractions have been used: S etching of valleys; L etching of grooves; R etching of deep holes, including vias

Solid	Parent/Carrier	Laser λ [nm]	W [μm/s] (I [W/cm^2]; ϕ [J/cm^2]) d [μm]	Remarks	Ref.
Ag	Cl_2	N_2	$W = 3$ Å/p (0.12)	p, QCM, AES, XPS, MS	[8.1, 2]
Ag	Cl_2	355, 532 Nd:YAG		P, AES, XPS	[8.3]
Al	Cl_2	XeCl, N_2	$W = 30$ Å/p (0.3); $W = 11$ Å/p (0.12)	λ, p, QCM	[8.2]
Al	Cl_2	XeCl	$W \leqslant 1.5$ μm/p (0.2)	S, HS, P, p, SEM; Al/Si	[8.4, 5]
Al	$HNO_3 + H_3PO_4 + K_2Cr_2O_7/H_2O$	Ar	$W = 1$; $d < 2$	L, EM 10^6, AES	[8.6]
Al	$Zn(CH_3)_2$; $C_2H_4O_2/H_2O$	257 + 515 Ar	(5 + 5E5)	Two-step microalloying etching	[8.7]
Al_2O_3/TiC (ceramic)	CF_4, CF_3Cl; CCl_4; SF_6	XeCl; Ar; Nd:YAG	$W = 50$ Å/p; $W = 0.1$; $W = 50$ Å/p		[8.8, 9]
Al_2O_3/TiC (ceramic)	KOH	Ar	$W = 200$ (1E6)	S, L, P, SEM	[8.10 – 12]
Au	Cl_2	N_2	$W < 6E-5$ (0.12)	S, QCM	[8.2]
$BaTiO_3$	H_2	Kr	$W = 200$	S, L, P, λ, SEM	[8.13, 14]
C	He	KrF; XeF	(2.5); (10)	UV, VIS	[8.15]
C	H_2	Hg arc lamp		HS, λ, K	[8.16]
CdS	$H_2SO_4 + H_2O_2/H_2O$	Ar		LEE, λ, SEM	[8.17]
CdS	$HCl + HNO_3/H_2O$	458, 514 Ar		LEE, λ, SEM	[8.18]
CdS	KCl, KBr, KI/H_2O	422 HeCd		LEE; gratings, diffraction efficiency	[8.19]
CdTe	–	Ar	$W = 50$	T; sputtering	[8.20]
Cu	Cl_2	355, 532 p-Nd YAG	($\leqslant 0.66$)	UV, VIS, QCM, AES, RBS, ESCA, X, MS	[8.89]

Table 8.1 (continued)

Solid	Parent/Carrier	Laser λ [nm]	W [μm/s] (I [W/cm²]; ϕ [J/cm²]) d [μm]	Remarks	Ref.
Cu	HCl/H_2O, $Br_2 + KBr/H_2O$	532 Nd:YAG XeCl, XeF	(<0.15)	S, L; CuBr particulates	[8.21, 22]
CuCl	Cl_2	355, 532 p-Nd:YAG	(≤1.3)	UV, VIS, QCM, AES, RBS, ESCA, X, MS	[8.89]
Fe	Cl_2	N_2	$W = 0.5$ Å/p (0.12)	S, QCM	[8.1, 2]
Fe_xNi_y	SF_6, CF_4, CCl_4	Ar	$W = 0.1$	QCM, XPS	[8.1, 8, 9]
Fe (stainless steel)	$NiCl_2/H_2O$	Ar	$W = 10$	S, LEE	[8.10, 11, 23, 24]
Fe-garnet	H_3PO_4	581 dye/Ar		$(BiGdLu)_3(FeGa)_5O_{12}$	[8.25]
GaAs	CF_3Br, CH_3Br, HBr	ArF	$W = 0.01$ (≤0.35) $d<1$	⊥, ∥, HS 330, P, K, SEM, LIF; (100), (111) n-GaAs, (100) p-GaAs, projection etching	[8.26–28]
GaAs	HCl/He	ArF	$W = 4E-4$	HS 300–500, XPS	[8.29, 30]
GaAs	CF_3I, CH_3Br, CH_3Cl	257 Ar	$W = 8E-3$ $d ≈ 1$	P, p; (100) n-GaAs	[8.31]
GaAs	Cl_2	488, 515 MLAr	$W = 33$	S, P, λ, p, SEM	[8.32]
GaAs	Cl_2 Br_2 CCl_4 $SiCl_4$	515 Ar	$W = 1$ $W = 0.1$ (5E4) $W = 6.5$ (4E6) $W = 2.4$ (5E5)	L, P, λ, p, T, RA, SEM, AES, X; (100) n-GaAs, tensile stress, LCVD of C in L	[8.33–36]
GaAs	CCl_4/H_2	532 Nd:YAG	$W = 0.14$	S, EM 10, P, p, SEM; 100 GaAs: Si, laser-enhanced RIE	[8.37]
GaAs	$H_2SO_4 + H_2O_2/H_2O$, HNO_3/H_2O, KOH/H_2O	257, 514 Ar, 257 SH, 488, 515, 633 Ar, 1.15 μm HeNe	(≤1E3) $W = 0.2$ $d = 1$	S, P, λ, SEM (100), (110), (111A), (111B) n-, p-, Si-GaAs	[8.17, 38, 96]
GaAs	$H_2SO_4 + H_2O_2/H_2O$	Hg lamp		P, n, p-GaAs	[8.39]

GaAs	$H_2SO_4 + H_2O_2/H_2O$	458, 515 Ar	(0.3)	SEM; gratings, different spacings	[8.40, 41]
GaAs	$KI + I_2/H_2O$	458 Ar	(4)	SEM; gratings	[8.42]
GaAs	$H_2SO_4 + H_2O_2/H_2O$ $H_2PO_4 + H_2O_2/H_2O$ KOH/H_2O	Ar	$W = 2$ $(1E-3)$ $d = 30$	S; (100) GaAs:Cr, Te, Si	[8.43]
GaAs	$KOH, HNO_3/H_2O$	334 – 515 Ar	$W = 10$ $(\leqslant 8E4)$	P, λ; (100) GaAs:Cr	[8.44, 45]
GaAs	$H_2SO_4 + H_2O_2/H_2O$	532 Nd:YAG		(100) GaAs:Si; surface ripples	[8.46]
GaAs	NaBr, NaI, KBr, KI, CsBr, CsI, $+ Br_2, I_2/H_2O$	413, 521 Kr, 633 HeNe	$W = 0.03$ $d = 70$	S, P, λ, SEM; gratings	[8.47]
GaAs	$KOH, HCl/H_2O$, $H_2SO_4 + H_2O_2/H_2O$	633 HeNe	(5E4)	L, SEM; n-GaAs	[8.48]
GaAs	$KOH/H_2O, C_2H_5OH$	633 HeNe	$W = 22$ (1E4)	P; n-GaAs	[8.62]
GaAs	$H_2SO_4 + NaSCN/H_2O$, $NaOH/H_2O$	633 HeNe	$W = 2E-3$ (0.75)	LEE; (111) p-GaAs:Zn	[8.49]
GaAs	Zn	647 Kr	$W = 0.012$ (200)	S, HS 900 – 1100, P	[8.50]
$GaAs_{1-x}P_x$	HCl/He	515 Ar; 766 dye/Ar	$W = 2E-4$ (500)	S, HS 380, LPE, P; $x = 0$; 0.02; 0.37; plasma	[8.51, 52]
$Ga_xAl_{1-x}As$	$H_3PO_4 + H_2O_2$ $+ CH_3OH/H_2O$	647 Kr	$d = 5$ (1)	L, VIS, TEM; heterostructures	[8.53, 54]
$Ga_{0.47}In_{0.53}As$	KOH/H_2O, KOH/C_2H_5OH	633 HeNe	$W = 1.2$ (1E4) $W = 7$ (1E4)	P, ρ	[8.62]
Ge	Br_2	Ar	$W = 1E-3$	$\parallel, P$	[8.55]
Ge	Br_2	488 dye/XeCl	$W = 0.2$ Å/p $(\leqslant 0.3)$	S, K, MS, TOF; (111) Ge	[8.56, 57]
Glass[a]	H_2	ArF	$W = 0.15$ μm/p $(\leqslant 2)$	Pyrex, projection patterning	[8.58]
Glass[a]	CF_2Br_2	ArF, KrF	$W = 0.5$ Å/p	S, K, UV, VIS, IR; Corning glass (70 SiO_2, 25 B_2O_3)	[8.59]
Glass[a]	HF	CO_2	$W = 0.025$ (1E3)	S; BK-7 glass[a]	[8.60]
InP	$CF_3I, CH_3Br,$ CH_3Cl	257 Ar	$W = 1E-3$	SEM	[8.31]
InP	CCl_4	515 Ar	$d \approx 1$		[8.36]

165

Table 8.1 (continued)

Solid	Parent/Carrier	Laser λ [nm]	W [μm/s] (I [W/cm^2]; ϕ [J/cm^2]) d [μm]	Remarks	Ref.
InP	$HCl + HNO_3/H_2O$	Ar			[8.17]
InP	H_3PO_4/H_2O	515 Ar	$W = 30$ (3E5)	L, P; (100) n-, p-, SI-InP	[8.61]
InP	AZ-303	458 Ar		SEM; gratings, diffractive efficiency	[8.42]
InP	$HNO_3 + HCl/H_2O$, $FeCl_3/H_2O$ KOH/H_2O KOH/C_2H_5OH	633 HeNe	— — $W = 0.2$ (1E4) $W = 1$ (1E4)	L, P, SEM	[8.62]
MnZn-ferrite	CCl_4, CF_4, SF_6, CF_3Cl	Ar	$W = 10$		[8.8, 9, 63, 64]
MnZn-ferrite	CCl_4	515 Ar	$W = 2.7$ (4E4)	P, SEM, AES; c-Fe_2O_3:MnO:ZnO $= 52:31:17$	[8.90]
Mo	NF_3	ArF	$W = 0.3$ Å/p (0.06) $d = 0.3$	SEM; lithography	[8.65 – 67]
Mo	Cl_2	N_2	$W = 3$ Å/p (0.12)	QCM	[8.2]
Ni	Cl_2	N_2	$W = 0.006$ Å/p (0.12)	S, QCM	[8.2]
Ni_xFe_{1-x}	Cl_2	N_2	$W = 0.7$ Å/p ($x = 0.8$) (0.12) $W = 0.2$ Å/p ($x = 0.47$) (0.12) $W = 0.4$ Å/p ($x = 0.2$) (0.12)	S, QCM	[8.2]
$Ni_{0.8}Fe_{0.2}$	SF_6, CF_4, CCl_4, Cl_2	Ar	$W = 0.1$		[8.8, 9]
Ni	$NiSO_4/H_2O$	Ar		EM10^4; electroless	[8.10, 11, 23, 24, 68]
$PbTi_{1-x}Zr_xO_3$ (PZT, PXE)	H_2	Ar, Kr	$W = 250$ (2E6)	S, L, P, λ, p, T, DK, SEM	[8.14]

166

Material	Etchant	Laser	Parameters	Notes / Diagnostics	Ref.
Si	COF₂/He	ArF	$W = 0.13$ Å/p (0.085) $d = 0.5$	L, P, SEM; poly-Si, high selectivity to SiO₂; lithography	[8.65 – 67]
Si	Cl₂	XeCl, Hg-Xe lamp	$W = 3.3E-3$ $W \leqslant 1.3E-3$ (0.28)	⊥, ∥, L, SEM; (100), (111) n-, p-Si, SI	[8.69 – 71]
Si	Cl₂	N₂	$W = 0.9$ Å/p (0.12) $W = 0.6$ Å/p (0.12) $W = 0.4$ Å/p (0.12)	a-Si (111) n-Si (0.1 Ω) (111) p-Si (0.1 Ω)	[8.1, 2]
Si	Cl₂, HCl	ML Ar	$W \leqslant 4E3$ $W \leqslant 1E3$ $d = 1$	L, P, λ, p, SEM; (100), (111)-Si	[8.72, 73]
Si	Br₂	Ar	$W = 1E-4$	∥	[8.74]
Si	CF₄ + O₂	515 Ar	$W = 8.3E-3$ (≤1E5)	LPE; n-, p-Si	[8.75]
Si	XeF₂	ML Ar	$W = 6E-4$	HS 300 – 630, P, p, K, MS; (111) n-, p-Si	[8.76 – 78]
Si	XeF₂	532 Nd:YAG p-CO₂	$W = 0.6$ Å/p (1)	P, λ, p, T, K, QCM, XPS, MS; SiF$_x$ (x≤3)	[8.1, 79 – 81]
Si	SF₆	p-CO₂	$W = 1.2$ Å/p ⊥ (1) $W = 0.05$ Å/p ∥ (≥10)	⊥, ∥, P, λ, p, K, IR, QCM, XPS, ESCA, MS; Ar, CF₄ buffer gases	[8.63, 64, 79, 80, 82]
Si	KOH/H₂O	Ar	$W = 15$ (1E7)	S, L, P, SEM; (100), (111)-Si	[8.10 – 12]
Si	HF/H₂O	515 Ar		Gratings	[8.83]
Si		Ar	$W = 0.5$ (1E5) $d = 40$	S, L; (111) Si	[8.43]
Si	NaOH/H₂O	Nd:YAG CO₂	$W = 0.05$ $W = 0.017$ (1E4)	S	[8.84]
Si	KOH/H₂O	Ruby	$W = 200$ Å/p $d = 0.4$	Etching and annealing	[8.85]
SiO₂	H₂	ArF		Projection patterning	[8.58]
SiO₂	NF₃/H₂	ArF	$W = 0.02$ Å/p (<8E–3)	⊥, ∥, L, K, IR, SEM, XPS	[8.29, 30]
SiO₂	CF₂Cl₂	KrF	$W = 0.13$ Å/p		[8.65 – 67]
SiO₂	CF₂Br₂, C₂F₄	KrF	$W = 0.6$ Å/p		[8.59]
SiO₂	Cl₂	458 Ar	$W = 3E-4$ (2E6) $d = 50$	S, λ, QCM	[8.79, 80]

Table 8.1 (continued)

Solid	Parent/Carrier	Laser λ [nm]	W [μm/s] (I [W/cm^2]; ϕ [J/cm^2]) d [μm]	Remarks	Ref.
SiO_2	Cl_2	Ar			[8.73]
SiO_2	CF_3Br, CDF_3	p-CO_2	$W = 0.3$ Å/p	$\parallel$, SEM, XPS, ESCA; MPD	[8.86, 87]
SiO_2	H_2SO_4/H_2O	532 Nd:YAG	(<0.15)		[8.21, 22]
$SrTiO_3$	H_2	Kr	$W = 200$	L, λ, SEM	[8.13]
Ta	SF_6, XeF_2	p-CO_2	$W = 2.6$ Å/p $W = 0.4$ Å/p	S, QCM	[8.63, 64, 79, 80]
Te	XeF_2	p-CO_2	$W = 10$ Å/p (1)	S, QCM	[8.79, 80]
Ti	NF_3	ArF	$W = 0.3$ Å/p	S, QCM	[8.65–67]
W	Cl_2	N_2	$W = 1.3E-3$ (0.12)	S, QCM	[8.2]
W	I_2			High selectivity to Si	[8.88]
W	COF_2/He	ArF	$W = 0.5$ Å/p	S, P, SEM	[8.65–67]
W	Air	ML Ar	$W = 12$ (4E6) $d = 21$		[8.95]
ZnSe	$HCl + HNO_3/H_2O$	458, 472, 488 Ar		LEE, λ, SEM	[8.18]

[a] Glass of complex composition or unspecified. Quartz glass, for example, is denoted by SiO_2, and crystalline quartz by c-SiO_2.

(Sect.3.5) are achieved at normal incidence of the laser light. Large-area etching can be performed at parallel incidence. Under otherwise identical experimental conditions (laser power and wavelength, kind and density of precursor molecules, optical configuration, etc.) the etch rate may strongly depend on crystal orientation and on the type and concentration of admixtures, impurities or dopants.

The most important dry-etching techniques currently used in micromechanics and microelectronics are plasma-assisted etching (PE) and reactive ion etching (RIE). Basically, PE and RIE involve reactive radicals and charged particles which interact with the solid surface, which is commonly held at or near ambient temperatures. In laser-induced etching the physical mechanisms involved are quite different and some fundamentals have already been discussed in Chaps.2 and 3. Therefore, laser-induced etching is a new technique that not only provides direct maskless etching at high rates but also makes it possible to process a wide variety of materials that cannot be processed by standard techniques, or only very inefficiently (Chap.9). The general trend in semiconductor microfabrication is definitely toward dry-etching, because wet-etching, in general, introduces high levels of contamination. Nevertheless, wet-etching is still of great importance, mainly due to the high rates and the great versatility that can be achieved with this processing technique. The greater versatility is related to the larger variety of reactants available without restrictions on volatility, etc. Laser-induced wet-etching may therefore become a useful tool in micromachining (cutting, drilling, etc.).

In the following we shall discuss dry-etching and wet-etching separately for metals, semiconductors, inorganic insulators and organic polymers. Dry-etching of metals and semiconductors is often classified into spontaneous etching systems, passive reaction systems and diffusive reaction systems ⌊8.1,2⌋. These terms will be explained below.

8.1 Metals

One of the best-known examples of abrasive reactive pyrolytic laser processing is the drilling and cutting of steel. Here, high power lasers are used to produce energy densities that bring about melting and vaporization of the material. When oxygen is used instead of air, cutting and drilling velocities can be remarkably increased due to the increase in the overall exothermal reaction energy. This technique is widely used and has been extensively reviewed ⌊8.90-93⌋. The lateral dimensions of holes and cuts are typically of the order of one to several millimeters. The axial dimensions range from several millimeters up to centimeters. Correspondingly, the mass of material removed is very large. In the following we concentrate on techniques that are more appropriate for applications in micromechanics and microelectronics. Here, the size of the workpieces and the volumes to be etched are generally much smaller and can therefore be treated in a sealed or flushed reaction chamber as described in Chap.4.

a) Dry-Etching

Laser-enhanced dry-etching of metals has been studied for Ag, Al, Au, Cu, Fe, Mo, Ni, Ta, Ti, W and some compounds of those (Table 8.1). The most-detailed investigations have been performed by CHUANG et al. [8.1-3,8,9,89,94].

Spontaneous etching occurs, for example, for Al in Cl_2 and for Mo, Ta, Ti, W in a XeF_2 atmosphere. This can be seen in Fig.8.1 for the example of the Al-Cl_2 system. After exposure of the Al film to a Cl_2 atmosphere, the frequency of the quartz crystal microbalance (QCM; see Sect.4.3) increases continuously with time (curve a). Irradiation with N_2 laser light increases

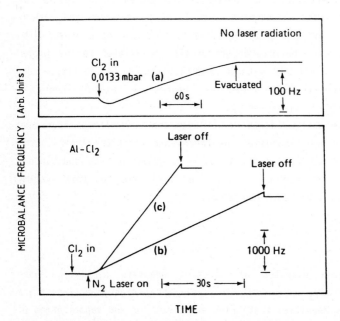

Fig.8.1. Frequency responses of a quartz crystal microbalance (QCM) covered with a thin film of Al exposed to a Cl_2 atmosphere. (a) $p(Cl_2)$ = 0.013 mbar, no laser irradiation. (b) $p(Cl_2)$ = 0.013 mbar, N_2 laser irradiation (0.12 J/cm^2, 30 pps). (c) $p(Cl_2)$ = 0.13 mbar, N_2 laser irradiation as in (b) (after [8.2])

170

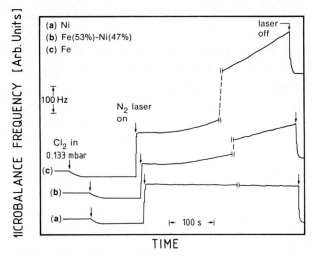

Fig.8.2. Frequency responses of quartz crystal microbalances covered with thin metal films exposed to a 0.13 mbar Cl_2 atmosphere and N_2 laser irradiation (12 MW/cm^2, 30 pps). (a) Pure Ni. (b) Fe$_{0.53}$Ni$_{0.47}$ alloy. (c) Pure Fe (after [8.1])

the etch rate further, depending on the laser fluence, and the Cl_2 pressure (curves b and c). For ϕ = 0.12 J/cm^2 and $p(Cl_2)$ = 0.13 mbar a rate of 11 Å/pulse has been obtained. Similar experiments using XeCl excimer laser radiation [ϕ = 0.3 J/cm^2, $p(Cl_2)$ = 1.3 mbar] yielded about 30 Å/pulse [8.2].

The class of *passive* reaction systems includes Au, Fe, Mo, Ni, W, and Ni_xFe_y alloys exposed to a Cl_2 atmosphere. The chlorine can chemisorb on such metal surfaces. This results in a surface reaction which stops after the formation of a passivated layer with a thickness of, typically, 10-30 Å. This thickness can be estimated from the initial decrease in microbalance frequency observed directly after Cl_2 exposure, which can be seen in Fig.8.2 for the example of Ni and Ni_xFe_y alloys. Irradiation with N_2 laser light at normal incidence results in a local temperature rise, which is indicated by the step-like increase in QCM frequency. The continuous increase in QCM frequency is related to the material removal that continuously increases with laser beam illumination time. While the rate of etching is almost negligible for pure Ni, it increases strongly with Fe alloying. This behavior shows the importance of the volatilities of solid reaction products in the etching process. $FeCl_3$ has a relatively high vapor pressure while chlorides of Ni are essentially nonvolatile below about 500 K. In other words, Ni chlorides will cover the solid surface and they will thereby suppress etching. The detailed microscopic mechanisms, for example the photon-enhanced desorption of

171

products by thermal and/or electronic excitation, are still under investigation.

Model systems for *diffusive* etching are Ag and Cu with Cl_2 [8.1-3,89]. In this case, chlorine not only chemisorbs on the metal surface, but also diffuses into the bulk material. This has been concluded from QCM, XPS and AES experiments. The chlorinated surface layer consists of $MeCl_x$, where x depends on the Cl_2 pressure and the time of exposure. For the example of Ag with 0.13 mbar Cl_2 and exposure times t ranging from 10 s ≤ t ≤ 1000 s the corresponding values of x were 0.67 ≤ x ≤ 0.95. The evidence for the formation of AgCl within the surface region was obtained from the observation of chemical shifts (about 2 eV), which are related to the metal-chlorine bonding. The exposure of the chlorinated metal surface to UV or VIS laser light results in efficient etching. Here, the etch rate depends not only on the laser power, wavelength and repetition rate but also on the Cl_2 pressure. The latter dependence is shown in Fig.8.3 for the example of the Ag-Cl_2 system and N_2 laser irradiation. As a result of the detailed investigations on the Ag-Cl_2 and Cu-Cl_2 systems by pulsed laser irradiation, the etching process can be described by a number of consecutive steps:

- Surface adsorption and diffusion of chlorine into the bulk
- Breaking or weakening of Me-Me bonds due to Me-Cl bonding
- Laser-excitation of the $MeCl_x$ layer via electronic excitation and surface heating
- Etching of the surface by desorption and ejection of Me, Cl and Me_yCl_x by the laser pulse.

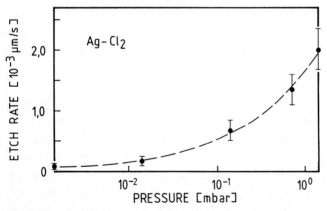

Fig.8.3. Etch rate of a Ag film for different Cl_2 pressures and N_2 laser irradiation (12 MW/cm^2, 30 pps) (after [8.1])

b) Wet-Etching

Laser-enhanced wet-etching of metals has been demonstrated for Al, Cu, Fe and Ni (Table 8.1). Pyrolytic wet-etching induced by Ar^+ laser radiation has been studied for thin films of Al in an aqueous solution consisting of HNO_3, H_3PO_4 and $K_2Cr_2O_7$ [8.6], with etch rates greater than 1 μm/s being obtained. This value exceeds the rate for dark etching by more than six orders of magnitude. The spatial resolution was better than 2 μm. Similar experiments have been performed by DONOHUE [8.21,22] for Cu in aqueous HCl. The mechanisms involved in those reactions are probably closely related to those described by VON GUTFELD et al. [8.10,11,23,24] for electroless etching (see below).

Laser-enhanced electrochemical etching (LEE) can be achieved by simply reversing the polarity of the cathode and anode in the experiments described for laser-enhanced plating (Sect.5.2.3). The mechanisms involved have been outlined in Sect.3.3. This etching technique has been used to produce 50 μm diameter holes in stainless steel using aqueous $NiCl_2$ as an electrolyte and Ar^+ laser radiation [8.10,11,23,24]. Etch rates of up to 10 μm/s have been obtained.

Electroless thermobattery or laser-enhanced exchange etching occurs for electrolytes possessing a negative temperature shift in the rest potential. The region illuminated by the laser light is etched and the peripheral region is plated. This was observed for a Ni-coated substrate in conjunction with an aqueous solution of $NiSO_4$ [8.10,11,23,24].

Wet-etching of Cu based on photolysis of the etchant has been described by DONOHUE [8.21,22]. The etchant was an aqueous solution of Br_2, sometimes with an admixture of KBr. XeCl and XeF excimer laser radiation was used to photodissociate the Br_2 (Sect.2.2.1). The bromine radicals can react with the Cu to form $CuBr_2$ and CuBr. While $CuBr_2$ is very soluble in water, CuBr is rather insoluble and may give rise to particulates that occur mainly at the edges of the illuminated area.

8.2 Semiconductors

Laser-induced chemical etching of semiconductors has been demonstrated for a wide variety of systems (Table 8.1). These include dry- and wet-etching of elemental semiconductors as well as III-V and II-VI compound semiconductors. The following discussion will concentrate on some model systems that have been extensively studied. Besides thermal activation of the etching reaction, electron-hole pair generation within the semiconductor surface, and selective electronic excitation (Si-Cl$_2$ system) or selective vibrational excitation (Si-SF$_6$ system) of the etchant will play an important role in the interpretation of the results.

8.2.1 Dry-Etching

The most-detailed investigations on laser-induced dry etching of semiconductors have been performed for Si and GaAs in halogen-containing gases such as Cl$_2$, XeF$_2$, SF$_6$, CCl$_4$ and CF$_3$Br.

a) The Si-Cl$_2$ System

High resolution etching of Si in Cl$_2$ by Ar$^+$ laser radiation has been investigated by EHRLICH et al. [8.73]. The Cl$_2$ pressures used ranged from 1.3 to 650 mbar. Figure 8.4 shows a scanning electron micrograph of etched grooves. The spatial resolution achieved at laser powers which induce melting of the Si surface is better than 1 µm. As outlined in Sect.2.2.1, Ar$^+$ laser radiation at wavelengths < 500 nm can photodissociate Cl$_2$ molecules according to (2.23). The Cl radicals strongly chemisorb on and diffuse into (see below) Si surfaces and thereby break Si-Si bonds. The etch rate depends strongly on the crystal orientation, the gas pressure, and the laser intensity and wavelength. Figure 8.5 shows the dependence of the etch rate (derived from

Fig.8.4. Submicrometer grooves etched with Ar$^+$ laser radiation into c-Si in a Cl$_2$ atmosphere (after [8.72])

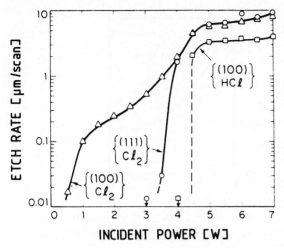

Fig.8.5. Etch rate for c-Si in Cl_2 and HCl [$p(Cl_2,HCl)$ = 266 mbar] atmospheres as a function of Ar^+ (multiline) laser power. The scanning velocity was v_s = 27 μm/s (after [8.73])

the etched depth of grooves) on the laser power for (100) and (111) Si-wafers. At low laser powers corresponding to temperatures well below the melting point, the etch rate for the (100) surface exceeds that for the (111) surface by at least 2 orders of magnitude. At higher laser powers, the rate becomes independent of crystal orientation when melting occurs, and also levels off in the same region. The latter behavior could be caused by the latent heat of melting and/or the step-like increase in reflectivity at the melting point. Additionally, mass transport limitations should also become effective at these laser powers and at the parameters used in this experiment (Sect.3.2). The influence of the latter mechanism could be derived from the dependence of the etch rate on laser focus diameter. The etching of Si under similar experimental conditions has also been investigated for other precursor molecules such as HCl and H_2. These molecules cannot be photodissociated with visible Ar^+ laser radiation. The etch rates achieved in the high power region, i.e. above melting of the Si surface, were about a factor of 2 to 3 smaller than those with Cl_2 (see Fig.8.5).

The complex nonlinear increase in etch rate in the low laser power region of Fig.8.5 directly shows that the (linear) photolysis of gaseous Cl_2 is not the only rate-determining mechanism. An explanation of this complex etching behavior requires consideration of additional light-molecule-surface interaction mechanisms. Among these are light-induced electron-hole pair generation within the Si surface, and local substrate-heating. Changes in

175

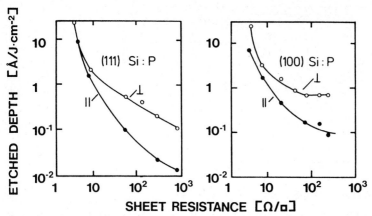

Fig.8.6. Etched depth as a function of sheet resistance of P-doped (111) and (100) Si in a 133 mbar Cl_2 atmosphere for perpendicular ($\perp$) and for parallel ($\parallel$) XeCl laser irradiation (after ⌊8.71⌋)

molecule-surface interactions will occur as a consequence of each of these, influencing adsorption, bond-breaking desorption of reaction products, etc.

The effect of light-induced electron-hole pair generation may be seen from Fig.8.6, which compares etch rates for normal and parallel XeCl laser irradiation of P-doped (n-type) (111) and (100) Si ⌊8.69-71⌋. For low doping concentrations (high sheet resistances), the number of electrons in the conduction band is essentially determined by the number of photoelectrons generated by the incident laser light. In this region, the etch rate is much higher for normal incidence than for parallel incidence. This difference could originate from an enhanced rate of Si-Cl covalent bond formation and - similarly to the Si-F reaction (see below) - an enhanced diffusion of Cl^- into the subsurface of the Si. Additionally, it is known that laser irradiation of semiconductors may produce surface electric fields as high as 10^{-4} - 10^{-1} V/Å, which are related to the different mobilities of electrons and holes (⌊8.99⌋; see also Sect.3.3). Such a surface electric field could strongly affect the surface chemisorption and diffusion of species into the lattice. In *highly* doped samples with many electrons in the conduction band, i.e. for low sheet resistances shown in Fig.8.6, the etch rate is nearly independent of the irradiation geometry and is essentially determined by the spontaneous etching being performed by chlorine radicals.

Figure 8.7 shows the etch rate for P- and B-doped Si as a function of sheet resistance. It becomes evident from the figure that the etch rate of both (100) and (111) Si strongly increases the more electrons there are in the conduction band. These results can again be tentatively explained by

176

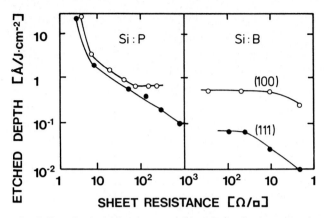

Fig.8.7. Etched depth as a function of sheet resistance for P- and B-doped (100) and (111) Si in a 133 mbar Cl_2 atmosphere under perpendicular XeCl laser irradiation (after ⌊8.71⌋)

surface-field-enhanced diffusion of Cl^- ions into the Si surface, in a similar way to that described above. Apparently, Cl^- ions can more easily penetrate the n-type surface than the p-type surface, where the electrons are the minority carriers. The influence of the etch rate on the crystallographic orientation may be explained by the diffusion coefficient of species, which is smaller for the closely packed (111) surface than for the (100) surface.

The results shown in Figs.8.6 and 7 represent the only experimental investigations of this type so far. They are in qualitative agreement with investigations on the Si-F reaction and the models derived for reactive laser-induced halogen etching of Si that are discussed below. In particular, the field-assisted diffusion mechanism may be similar to that proposed by WINTERS et al. ⌊8.100,101⌋ for the Si-F reaction. According to this model the surface halogenation process in the Si etching reaction should be similar to that in photooxidation. It was proposed in Sect.6.1 that lattice disruption and Coulombic forces may strongly contribute to the photoenhancement of the Si oxidation rate. However, with respect to Figs.8.6 and 7, it should be emphasized that quantitative conclusions should not be drawn. One reason is that it is not clear from these experiments whether the diminution of the laser light within the Cl_2 atmosphere, which is very significant for the high gas pressures ⌊$p(Cl_2)$ = 133 mbar⌋ used, has been taken into account properly. Another point is that Cl-Cl radical recombination will also play a very important role, especially at parallel laser beam incidence.

For high laser irradiances that generate very high electron densities, Auger recombination can be extremely rapid and the light energy absorbed is directly converted into heat (Sect.2.1). In this regime the surface chemical process is mainly thermally activated. This may explain why no difference in etch rate was found for n-and p-type Si under focused Ar^+ laser irradiation ⌊8.73⌋. A similar effect has been observed in laser-induced oxidation of Si where at high laser irradiances quantum effects are washed out (see Sect.6.1).

The etching of p-Si and c-Si in Cl_2 atmosphere by UV Hg-Xe lamp irradiation has been investigated in detail by OKANO et al. ⌊8.70⌋.

b) The Si-XeF$_2$ System

Contrary to Cl_2, XeF_2 can etch Si spontaneously, i.e without any external radiation. The major volatile product is SiF_4, and to a minor extent SiF_x with x ⩽ 3. Variation of the Si temperature over the range 300-630 K does not change the distribution of the type of reaction products, but only increases the overall reaction rate. On the other hand, the distribution of reaction products may depend on the orientation of the Si surface ⌊8.102⌋. Synchrotron and XPS experiments on the Si-XeF$_2$ system have revealed that F atoms may chemisorb on the Si surface and, additionally, diffuse beneath the surface ⌊8.102⌋.

The enhancement of the etch rate and the distribution of reaction products under UV, visible and IR laser light irradiation has been studied in great detail. Most of the experiments have been performed with 514.5 nm Ar^+ and 10 μm CO_2 laser radiation, neither of which can excite gaseous XeF_2 in a single-photon process. HOULE ⌊8.76-78⌋ found that direct band-gap excitation of the Si by Ar^+ laser light influences both the etch rate and the distribution of reaction products. The etch rate increases more or less linearly with laser power (except for the lowest powers used). The major desorbed species are SiF_4 and SiF_3. The relative amount of SiF_4 to SiF_3 decreases with increasing laser power. Since the decrease in SiF_4 is matched by the increase in SiF_3, it seems reasonable to assume that SiF_3 is a precursor to SiF_4 prior to desorption. In the same way as in the Si-Cl_2 system, the etch rate for n-type Si was found to be higher than for p-type Si (see above and Sect.3.3). For pulsed laser irradiation, the light-molecule-surface interaction mechanisms seem to be somewhat different from those involved in the cw Ar^+ laser-enhanced reaction. This was discovered by CHUANG et al. ⌊8.1,79-81⌋. For example, pulsed irradiation with 532 nm frequency-doubled Nd:YAG laser light resulted in the desorption of SiF_x (x ⩽ 3) fragments and Si atoms. The yield for less F-coordinated species and Si atoms was found to increase with increasing laser intensity. Interestingly, sub-band-gap irradiation with pulsed CO_2 laser light also produced a product distribution quite different from the thermal reaction, i.e. the dominant species were again not SiF_4 but SiF_x with $x \leqslant 3$. It may be possible that transient thermal electrons (Sect.2.1.1) generated by single- and/or multiphoton processes create transient electric fields within the Si surface and thereby enhance the etch rate as described above.

c) The Si-SF_6 System

A model system in which the importance of selective multiphoton vibrational excitation has been clearly demonstrated is the etching of Si by SF_6 under CO_2 laser radiation at intensities $\leqslant 20$ MW/cm^2 (see Sect.2.2.3 and $\lfloor 8.8,9,63,64,82 \rfloor$). While Si is almost inert against SF_6, CO_2 laser irradiation can induce high etch rates, particularly at normal incidence. The major reaction products are SiF_4 and SF_4. The overall reaction can be divided into a number of consecutive steps: In the first step, SF_6 molecules in the gas or physisorbed phase are excited into high vibrational states via coherent multiphoton excitation according to (2.34). Subsequently, vibrationally excited SF_6^* can dissociatively chemisorb on the Si surface

$$SF_6^*(g) \longrightarrow SF_6^*(ads) \text{ or } SF_6(ads) \text{ ,} \tag{8.1}$$
$$SF_6^*(ads) \longrightarrow SF_4(g) + 2F(ads) \text{ .} \tag{8.2}$$

At very high laser light intensities $\lfloor$see below and (2.35)$\rfloor$, F radicals are produced. These can spontaneously chemisorb on Si surfaces

$$F(g) \longrightarrow F(ads) \text{ .} \tag{8.3}$$

As identified by XPS, chemisorption may lead to a SiF_2-like state

$$Si(s) + 2 F(ads) \longrightarrow SiF_2(ads) \text{ ,} \tag{8.4}$$

where F may also penetrate beneath the surface as in the Si-XeF_2 reaction. The reaction between chemisorbed species according to

$$SiF_2(ads) + 2 F(ads) \longrightarrow SiF_4(ads) \text{ ,} \tag{8.5}$$
$$SiF_2(ads) + SF_6^*(ads) \longrightarrow SF_4(g) + SiF_4(ads) \text{ ,} \tag{8.6}$$

results in the formation of SiF_4 that desorbs from the surface

$$SiF_4(ads) \longrightarrow SiF_4(g) \text{ .} \tag{8.7}$$

179

Clearly, these are only the main reaction pathways. Other products such as SiF_x ($x \leqslant 3$) may be formed to a minor extent. Additionally, at normal incidence the laser radiation may also affect the surface chemistry in various other ways, such as

$$SiF_2(ads) + SiF_2(ads) + nh\nu(CO_2) \longrightarrow SiF_4(ads) + Si \ , \qquad (8.8)$$
$$SiF_4(ads) + nh\nu(CO_2) \qquad\qquad \longrightarrow SiF_4(g) \ , \qquad (8.9)$$

where $n \geqslant 1$. The effect of direct substrate irradiation can be estimated by comparing reaction yields for normal and parallel incidence. The decrease in reaction yield observed for the parallel irradiation geometry cannot be explained by gas-phase collisional relaxation effects only. The major surface reaction steps in the chain of reactions outlined above include photon-enhanced dissociative chemisorption (8.2), reaction of vibrationally excited SF_6^* with the fluorinated surface (8.6) and desorption of SiF_4 (8.7). This seems to be true at least for low laser intensities where (2.34) dominates (2.35) as described in Sect.2.2.3. That selective nondissociative multiphoton vibrational excitation of the SF_6 is indeed the important mechanism in the low intensity range becomes evident from a number of observations: The etch rate shows a pronounced wavelength dependence with a maximum that is much broader and lower in frequency than the single-photon absorption spectrum of the ν_3 SF_6-mode at 948 cm^{-1} (Fig.8.8). The power dependence of the etch rate (Fig.8.9) can be described by $W \propto P^{3.5}$, which indicates that the overall

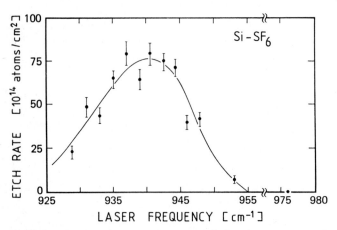

Fig.8.8. Etch rate for Si in SF_6 $\lfloor p(SF_6) = 3.3$ mbar$\rfloor$ as a function of CO_2 laser frequency ($\phi = 0.9$ J/cm^2). Laser light irradiation was perpendicular to the Si substrate. Each data point is an average over 15 laser pulses (after $\lfloor 8.82 \rfloor$)

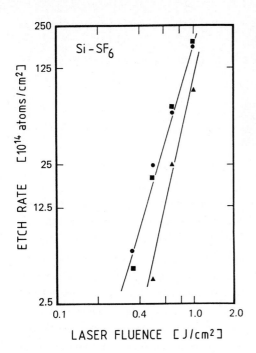

Fig.8.9. Etch rate for Si in SF_6 as a function of CO_2 laser fluence at $\nu = 942.4$ cm^{-1}, normal incidence and 30 laser pulses for each data point. ▲: $p(SF_6) = 1.1$ mbar; ●: 2.7 mbar; ■: 6.7 mbar (after [8.82])

Si - SF_6

ETCH RATE [10^{14} atoms/cm^2]

LASER FLUENCE [J/cm^2]

rate-limiting step is based on selective three- or four-photon excitations. The reaction yield increases monotonically with SF_6 gas pressure and saturates at about 2 mbar (see also Sect.2.2.3).

The main reaction mechanisms become different for very high light intensities (⩾ 50 MW/cm^2) that are produced by focusing the CO_2 laser light. At normal incidence, such intensities would evaporate the Si substrate, therefore, processing can only be performed with *parallel* incidence of the laser light (Fig.4.3). At these laser light intensities, SF_6 molecules are decomposed into SF_5 and F by multiphoton dissociation (MPD) according to (2.35). The unstable SF_5 further dissociates into SF_4 and another F atom; the F atoms diffuse to the solid surface and react with it according to (8.3) and the consecutive steps outlined above.

Clearly, MPD can be applied only in large-area etching. Nevertheless, this technique seems to be promising since many halogen-containing molecules, such as CF_3X, CF_2X_2 (X = Cl, Br, I), N_2F_4 etc., can be readily decomposed by MPD to produce reactive radicals for possible reactions not only with Si but for a wide variety of other solids.

HOLBER et al. [8.75] have described Ar$^+$ laser-enhanced plasma-etching of p- and n-type Si in a mixture of CF_4 and O_2. The enhancement in etch rate was found to be based on both pyrolytic and photolytic mechanisms.

Fig.8.10. Cross-sectional view of a groove etched in (100) GaAs in 85 mbar CCl_4 with 515 nm Ar^+ laser light (after ⌊8.33⌋)

d) Compound Semiconductors

Dry-etching of compound semiconductors has been mainly demonstrated for GaAs and, less extensively, for CdS and InP. The precursor molecules most commonly used were Cl_2, CCl_4, CF_3Br and CH_3Br.

TAKAI et al. ⌊8.33-35⌋ have studied Ar^+ laser-induced etching of (100) n-type GaAs under Cl_2, CCl_4 and $SiCl_4$ atmospheres. Figure 8.10 shows a scanning electron micrograph of a groove etched in 85 mbar CCl_4. A focus of 1.2 μm (λ = 514.5 nm) and a scanning speed of 9 μm/s were used. The etched width and depth of the line are about 1 μm and 0.25 μm, respectively. Figure 8.11 shows the width and depth of grooves as a function of scanning velocity for two

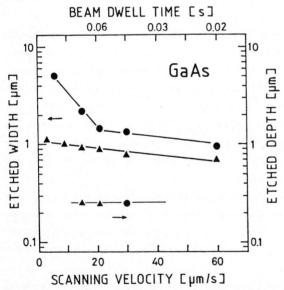

Fig.8.11. Width and depth of grooves etched with 515 nm Ar^+ laser light in GaAs as a function of scanning velocity. $p(CCl_4)$ = 87 mbar. •: P = 126 mW; ▲: P = 79 mW (after ⌊8.33⌋)

laser powers. The groovewidth decreases with decreasing laser power and increasing scanning velocity. For the 126 mW curve and a scanning velocity of 60 μm/s, corresponding to a dwell time of 0.02 s, the etch rate is estimated to be about 1.5 μm/s (see also Sects.3.2 and 5.2.2). The width of the grooves can be reduced to about 0.6 μm by lowering the CCl_4 pressure to 40 mbar (see Sect.3.5). For both Cl_2 and CCl_4, the etch rates were similar for multiline and 514.5 nm single-line Ar^+ laser operation. No etching was observed for CCl_4 and parallel light incidence at a distance of < 0.5 mm from the substrate. From these results and from the power dependence of the etch rate it was concluded that etching is primarily thermally controlled. This interpretation, however, should be verified still further. In particular, the power dependence of the etch rate would be similar if laser-induced thermal desorption of reaction products were rate limiting. Furthermore, electron-hole pair generation could modify the surface adsorption of and the reaction with chlorine considerably. The mechanisms could be similar to those described for Si. In fact, recent investigations by ASHBY [8.51] yield indications for the importance of nonthermal processes based on light-induced electron-hole pair generation. In these experiments a glow discharge was used to produce Cl radicals from an admixture of HCl and He. The concentration of Cl radicals produced was held so low that measurable etching of the GaAs could be observed only under simultaneous Ar^+ laser irradiation.

Large-area etching of (100) and (111) n-type GaAs and (100) p-type GaAs by means of ArF laser radiation has been reported by BREWER et al. [8.26-28]. CF_3Br, and to a minor extent also CH_3Br, served as precursor molecules. ArF laser radiation photodissociates CF_3Br and CH_3Br, thereby producing Br and CF_3 and CH_3 radicals, respectively (Sect.2.2.1). These radicals may react with the GaAs surface and form a large number of products that are detected mainly by laser-induced fluorescence (LIF). For low laser fluences, the etching process is mainly nonthermal. Evidence for this interpretation results from experiments using XeF excimer laser radiation in otherwise identical experimental conditions. XeF laser light does not photodissociate CF_3Br and CH_3Br but may induce the same substrate temperature. Under these conditions no etching was observed. Nevertheless, the etch rates observed with ArF laser radiation at normal incidence were higher than those for parallel incidence. This difference results, at least in part, from the laser-enhanced thermal desorption of nonvolatile etching products, which becomes effective only at normal incidence. Blocking of reactive surface sites by nonvolatile reaction products can also be diminished by uniform substrate heating. As expected, an exponential increase in etch rate with

183

substrate temperature was found. The etching process is strongly anisotropic. For GaAs crystal orientations (111B), (100) and (111A), the typical ratios of etching were 3:2:1. For energy densities < 35 mJ/cm^2 the surface morphology is relatively smooth. Etch rates up to about 10^{-2} μm/s were achieved. At energy densities > 35 mJ/cm^2, physical ablation seems to be the primary etching mechanism. For such fluences, however, the surface morphology becomes rough and damaging of the material is observed.

Large-area patterning of GaAs was demonstrated by both direct masking and projection of the masked excimer laser beam onto the substrate. Here, a resolution of about 0.2 μm was obtained.

Laser-enhanced reactive ion etching (RIE) of GaAs was demonstrated by TSUKADA et al. ⌊8.37⌋. Frequency-doubled Nd:YAG laser light and an atmosphere consisting of a mixture of CCl_4 and H_2 were used in these experiments. The etch rate was enhanced by at least a factor of ten. The most likely mechanism is laser-enhanced thermal desorption of the reaction products.

8.2.2 Wet-Etching

Laser-enhanced wet-etching based on pyrolytic and photolytic mechanisms has been investigated for element and compound semiconductors (Table 8.1). Etchants include aqueous solutions of HCl, H_2SO_4, HNO_3, KOH, NaOH, H_2O_2 and admixtures of these.

a) Silicon

Silicon immersed in aqueous KOH has been etched with focused Ar$^+$ laser radiation ⌊8.12⌋. Figure 8.12 shows the volume etch rate as a function of laser power for (111) Si and for ceramic Al_2O_3/TiC. Each data point represents the average of the volume of 10-20 holes produced with pulses of τ = 5 s at constant laser power (see also Sect.8.3). Average etch rates of up to 15 μm/s have been obtained. For the high laser powers, melting and/or vaporization dominate the etching process. This follows from the observation that above a certain threshold intensity the etch rates for (111) and (100) Si surfaces are similar ⌊at lower temperatures etch rates in (100)-direction may exceed those in (111)-direction by more than 2 orders of magnitude⌋. The high etch rates achieved become possible because of the significant increase in mass transport due to microstirring (Sects.3.3 and 5.2.3). It should be noted also that the etched features did not show any lips or melted and refrozen ridges.

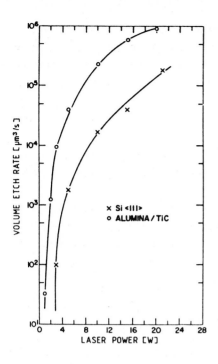

Fig.8.12. Volume etch rate for (111) Si and ceramic Al$_2$O$_3$/TiC in KOH as a function of the Ar$^+$ laser power. Full curves are guides for the eye (after [8.12])

b) Gallium Arsenide

At *low* light intensities, the microscopic mechanisms at the liquid-semiconductor interface become quite different from those described in the preceding paragraph. In this processing range, photo-generated carriers at the interface may induce or accelerate the etching. The mechanism in question has already been described in Sect.3.3. The most-detailed experiments have been performed for GaAs with both visible and UV laser radiation [8.17,38,40,41,44,45]. Similar to dry-etching, wet-etching is also strongly anisotropic. This has been demonstrated for different visible Ar$^+$ laser lines with an aqueous solution of H$_2$SO$_4$ and H$_2$O$_2$ (volume ratio H$_2$SO$_4$: H$_2$O$_2$: H$_2$O = 1:1.3:25) as etchant. While etch rates were similar for (100) and (110) surfaces, etching for (111A) surfaces was about five times faster. The dark rate for (111B) faces was so high that reliable measurements of the photoenhancement were impossible [8.17]. At low intensities, the etch rate for n-type GaAs (dopant concentration 10^{18} atoms/cm^3) was 5 times higher than for semiinsulating material. No etching was observed for p-type (dopant concentration 10^{18} atoms/cm^3) material. These differences in etch rates can be attributed to the differences in band bending described in Sect.3.3. Figure 8.13 shows etch rates as a function of laser intensity for n-type GaAs in different etchants and for two laser wavelengths [8.38]. The etching

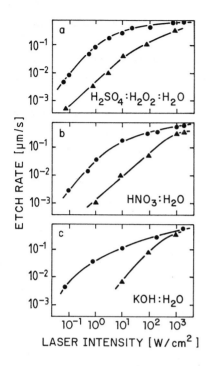

Fig.8.13a-c. Etch rates for n-type GaAs in various aqueous solutions as a function of the intensity of (▲) 515 nm Ar+ and (●) 257 nm frequency-doubled Ar+ laser radiation. The compositions of the etchants were (a) $H_2SO_4:H_2O_2:H_2O$ = 1:1:100 (by volume); (b) $HNO_3:H_2O$ = 1:20 (by volume); (c) $KOH:H_2O$ = 1:20 (by weight) (after ⌊8.38⌋)

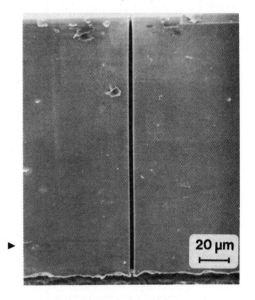

Fig.8.14. Scanning electron micrograph of a via hole etched through a 200 μm thick GaAs sample (after ⌊8.38⌋)

solutions were diluted in such a way that dark etching and UV activation of the bulk liquid were minimal or negligible. The different curves are characterized by an initial linear increase in rate and a saturation above about 10^3 W/cm². The saturation may originate from mass transport limitations. This can be concluded from the observation that the rate becomes independent of the laser wavelength and very similar for the three different etching solutions. Another remarkable feature is the difference in etch rates observed for UV and visible laser light at low powers. It has been suggested that the rapid etching under UV laser irradiation originates from hot, nonthermalized holes that are injected into the interface ⌊8.38⌋. In contrast to visible light, UV light also allows p-type materials to be etched, though at smaller rates, and at *low* laser intensities. At higher intensities, thermal effects will again influence the etching process.

Liquid-phase etching of GaAs makes it possible to produce deep, high-quality via holes with perfectly vertical walls. An example is shown in Fig.8.14. The enormous aspect ratio achieved can be attributed to waveguiding of the laser beam. This effect has already been observed in laser machining of metals (see e.g. ⌊8.90,92⌋).

Another interesting application of liquid-phase etching is maskless direct etching of submicrometer gratings. The most-detailed investigations have been performed for GaAs in an aqueous solution of H_2SO_4 and H_2O_2. Here, visible Ar^+ laser radiation was employed together with a non-right-angle corner consisting of a mirror on one side and the GaAs substrate on the other. The superposition of the incident laser beam and the beam reflected from the mirror produced an interference pattern on the GaAs surface. As a consequence, a periodic structure was etched. By varying the corner angle, different groove profiles, including blazed, sinusoidal, and "impulse" shapes, were produced. Examples are shown in Fig.8.15. The spatial resolution of the technique was about 200 nm. Depth-to-spacing ratios ranged from about 0.2 to 0.8.

Laser-enhanced *electrochemical* etching of GaAs has been described by OSTERMAYER and KOHL ⌊8.49⌋.

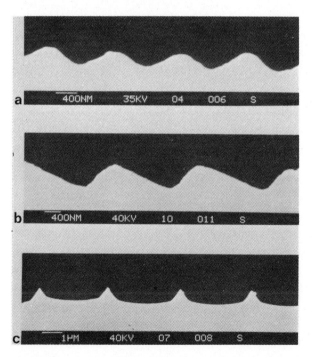

Fig.8.15a-c. Scanning electron micrographs of gratings with (a) sinusoidal, (b) blazed, and (c) impulse profiles (after ⌊8.40⌋)

8.3 Inorganic Insulators

Etching of inorganic insulators has been investigated mainly for amorphous SiO_2 (a-SiO_2; often simply denoted by SiO_2), for some glasses of complex composition, for oxidic perovskites such as $BaTiO_3$, $SrTiO_3$, $PbTi_{1-x}Zr_xO_3$, and for ceramic Al_2O_3/TiC (Table 8.1).

8.3.1 SiO_2 - Glasses

Photochemical dry-etching of SiO_2 and SiO_2-rich glasses has been investigated for various reactant molecules that have been dissociated either by single-photon electronic or by multiphoton vibrational excitations. For the molecules investigated, spontaneous etching is zero or negligible (Sect.8.1).

BRANNON [8.59] and LOPER and TABAT [8.65-67] used ArF and KrF excimer lasers to photodissociate CF_2Cl_2, CF_2Br_2 and CF_3Br, CF_3I, CF_3NO and $CO(CF_3)_2$. Interestingly, etching was very efficient for CF_2Cl_2 and CF_2Br_2 (typical etch rates were 0.2-0.5 Å/pulse), while the other compounds yielded either very inefficient etching or else none at all. These results seem to suggest that CF_2 radicals interact with the SiO_2 surface more strongly than CF_3 radicals. The reason for the different behavior of these radicals is not understood and the results need to be further verified. Experiments carried out under similar conditions, but with Br_2 and SF_6 as parent gases did not yield any appreciable etching. HIROSE et al. [8.29,30] have investigated the etching of SiO_2 in a mixture of NF_3 and H_2 under ArF excimer laser irradiation. The reaction mechanism suggested was qualitatively described by

$$SiO_2 + NF_3 + H_2 + h\nu(193 \text{ nm}) \longrightarrow SiF_4 + NO_2 + N_2O + HF . \qquad (8.10)$$

Here, the etch rate was found to increase with H_2 partial pressure. The main reaction products observed were SiF_4, N_2O and NO_2.

Etching of SiO_2 in a Cl_2 atmosphere by visible Ar^+ laser radiation has been investigated by CHUANG et al. [8.8,9,79,80]. Because SiO_2 is transparent within the visible spectral region, laser-induced excitation or heating of the substrate cannot be of great importance. In fact, the etch rate seems to be correlated with the photodissociation yield of the Cl_2 (Sect.2.2.1). For 457.9 nm radiation an etch rate of up to $3 \cdot 10^{-4} \mu m/s$ [$p(Cl_2)$ = 133 mbar] was obtained, while the rate was significantly lower for the 514.5 nm laser line, under otherwise identical experimental conditions. The lateral dimensions of etched features were much larger than the focal spot size. For example, for $2w_0 \approx 7 \mu m$ the diameter of the holes was 50-80 μm, depending on the laser beam illumination time. This can be explained by the random diffusion of Cl radicals produced within the gas phase (Sect.3.5).

Large-area etching of SiO_2 activated by multiphoton vibrational dissociation (MPD) of CF_3Br and CDF_3 with pulsed CO_2 laser radiation has been reported by STEINFELD et al. ⌊8.86,87⌋. Because of the high laser intensities necessary for MPD, such experiments can be performed only in the parallel configuration (Fig.4.3). The etch rate achieved with the laser focus 1 mm above the substrate was 0.3 Å/pulse. It was proposed that the CF_3 radicals generated in the MPD process react with SiO_2 to form volatile products. This interpretation is in disagreement with the results mentioned above. It remains unclear whether the different observations are related to the differences in optical configuration, photochemistry and/or concentrations of radicals generated in both experiments.

Projection etching of SiO_2-rich glasses by transient heating in a H_2 atmosphere has been demonstrated with ArF laser radiation [8.58]. Gratings with a resolution of about 0.4 μm have been produced. Figure 8.16 shows the etched depth as a function of laser fluence for Pyrex and thermally grown SiO_2 on Si.

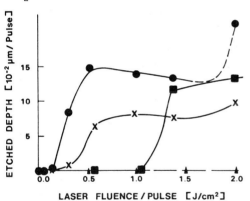

Fig.8.16. Etched depth of various different glasses in H_2 atmosphere as a function of ArF laser fluence. Pyrex: •: $p(H_2)$ = 266 mbar; x: $p(H_2)$ = 1333 mbar. SiO_2 on Si: ■: $p(H_2)$ = 266 mbar (after ⌊8.58⌋)

8.3.2 Perovskites

Laser-induced etching of oxidic perovskites has been investigated by EYETT et al. ⌊8.14,103-106⌋. Experiments were performed with visible and UV Ar^+ and Kr^+ laser radiation and with excimer laser radiation. As outlined in Sect.6.2, laser-induced heating of these materials in a reducing atmosphere results in the formation of oxygen vacancies and quasi-free electrons. With increasing laser powers, metallization and, finally, etching occur. The most-detailed investigations on etching have been performed for crystalline

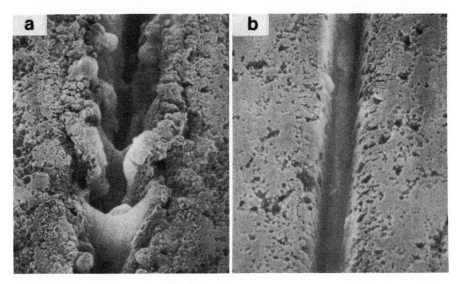

Fig.8.17a,b. Scanning electron micrographs of grooves etched with 647 nm
Kr$^+$ laser radiation in ceramic PbTi$_{1-x}$Zr$_x$O$_3$. The ambient medium was (a) air
with p(air) = 90 mbar, (b) H$_2$ with p(H$_2$) = 90 mbar. P = 0.72 W, 2w$_0$ = 18 μm,
v$_s$ = 8.4 μm/s (after [8.14])

BaTiO$_3$ and SrTiO$_3$ and for ceramic BaTiO$_3$ and PbTi$_{1-x}$Zr$_x$O$_3$ (PZT, PXE) in an
H$_2$ atmosphere. Scanning electron micrographs of grooves produced in PZT with
647 nm Kr$^+$ laser radiation in air and in an H$_2$ atmosphere under otherwise
identical processing conditions are compared in Figs.8.17a,b. While materials
removal in air is very irregular, well-defined structures can be produced in
H$_2$. Additionally, etching thresholds are considerably lower in H$_2$ than in
air. The role of H$_2$ is interpreted by the local decrease in O$_2$ partial
pressure, which favors efficient local reduction of the material (see
Sect.6.2). This has two consequences:
   - The local absorption increases strongly and becomes spatially better
defined.
   - The large number of oxygen vacancies generated favors a local collapse
of the perovskite lattice.
   Figure 8.18 shows the width and depth of grooves and holes as a function of
laser power. The illumination time for the holes was chosen to be equal to
the dwell time of the laser beam during etching of grooves, i.e. t$_d$ = 2w$_0$/v$_s$.
For low laser powers, the width is approximately equal to the diameter of the
laser focus. For the highest laser powers used, the width increases by a
factor of two to three. The etched depth increases continuously but shows a
decreasing slope with increasing laser power. The latter observation may

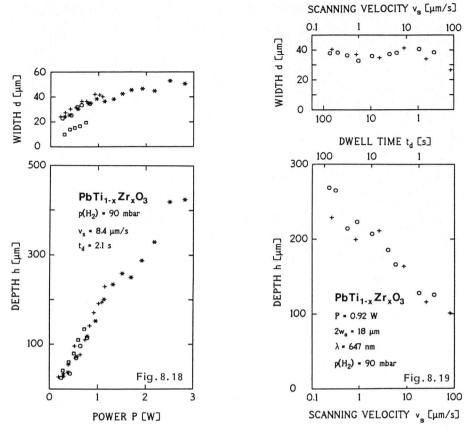

Fig.8.18. Width and depth of grooves and holes etched in ceramic PbTi$_{1-x}$Zr$_x$O$_3$ as a function of the incident Kr$^+$ laser power. +: grooves, 2w$_0$ = 18 μm, λ = 647 nm, v$_S$ = 8.4 μm/s. *: grooves, 2w$_0$ = 18 μm, λ = 488 nm, v$_S$ = 8.4 μm/s. □: grooves, 2w$_0$ = 11 μm, λ = 647 nm, v$_S$ = 8.4 μm/s. o: holes, 2w$_0$ = 18 μm, λ = 647 nm, t$_d$ = 2.1 s (after [8.14])

Fig.8.19. Width and depth of grooves and holes etched by 647 nm Kr$^+$ laser radiation in PbTi$_{1-x}$Zr$_x$O$_3$ in a H$_2$ atmosphere as a function of scanning velocity and laser beam dwell time. +: grooves; o: holes (after [8.14])

result from the diminution of the effective laser power by the material ejected out of the etched groove or hole. At constant power, the etched width stays about constant while the depth increases logarithmically with the dwell time of the laser beam (Fig.8.19). Temperature measurements using visual and photoelectric pyrometry (Sect.4.4) revealed an Arrhenius type behavior of the average etching depth. For temperatures ≤ 1600 K, an apparent activation energy of 41 ± 8 kcal/mole was derived. In this connection it seems interesting to note that the energies of evaporation of pure PbO [8.107] and

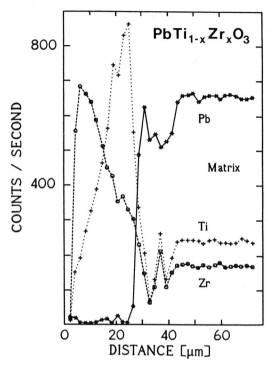

Fig.8.20. Electron-beam-induced X-ray fluorescence signal for Pb, Ti and Zr as a function of the distance from the groove edge. The measurements were made near the surface of the sample, on a V-shaped cross section similar to that shown in Fig.2.9c (after [8.14])

of PbO out of $PbTi_{1-x}Zr_xO_3$ ($x = 0.65$) [8.108] are 56 kcal/mole and 39 kcal/mole, respectively. The corresponding energies of evaporation of TiO_2 and ZrO_2 are 137 and 170 kcal/mole, respectively [8.107]. This is a hint that the evaporation of PbO is important in the etching process. This conclusion is supported by SEM and electron-beam-induced X-ray fluorescence studies. In a region of about 30-50 μm around the edges of grooves or holes, morphology changes are observed, which are due to a depletion of Pb (Fig.8.20). Beyond the depletion layer, the morphology and chemical composition corresponds to that of bulk PZT. The thickness of the depletion layer depends on the incident laser power. Results similar to those presented in Figs.8.18 and 19 were obtained for the other perovskites investigated.

The etching conditions became quite different, however, when UV laser radiation with an energy exceeding the band-gap energy, E_g, of the perovskite under consideration was used instead of visible laser radiation. Initial experiments were performed with the 337-356 nm multiple line output of the

Kr^+ laser. Up to the highest laser powers available (P ≤ 400 mW), the etch rates were in good agreement with those presented in Fig.8.18, but they were independent of the surrounding atmosphere (H_2 or air). These results can tentatively be explained by the strong increase in the absorption coefficient for photon energies $h\nu > E_g$. Furthermore, experiments with 308 nm XeCl laser radiation have shown that the depletion layer mentioned above can be considerably reduced or even avoided ⌊8.106⌋.

8.4 Organic Polymers

Laser-induced ablation of organic polymers and biological materials can be performed in a vacuum or in a nonreactive atmosphere. As already outlined in Sect.3.4, the process is mainly thermally controlled for IR and visible laser radiation, while direct nonthermal bond breaking seems to dominate for UV radiation below about 200 nm at low to medium laser irradiances. While this section deals mainly with the processing of organic polymers, the technique can be applied also to biological materials that have similar properties.

Table 8.2 summarizes the results currently available on the etching of organic polymers. It can be seen that most of the experiments have been performed in air. Systematic investigations on the influence of a reactive surrounding on the etch rates are still lacking.

Pyrolytic ablation of polymers and biological materials is based on local melting and/or evaporation. The local laser-induced temperature rise can be estimated from the equations presented in Sect.2.1.1. Due to their temperature sensitivity, organic materials become, in general, heavily damaged and/or distorted during pyrolytic processing. While this may be irrelevant in particular applications, such as in large-area cutting for example, the technique does not allow well-defined patterning in the sub-millimeter region. The following discussion will therefore concentrate on *photochemical ablation* (APD). This technique, which opens up a wide variety of new and quite different applications, not only in materials processing but also in medicine, was pioneered by SRINIVASAN and co-workers ⌊8.112-114,116,119,120,125,128-130⌋.

Figure 8.21 shows scanning electron micrographs of two polymers, PMMA and Riston (Du Pont Riston^{TM} is a negative photoresist, which consists of PMMA and other acrylates), that have been irradiated with ArF laser light through mechanical masks. The relatively sharp edges and the high aspect ratios achieved are typical of APD.

Table 8.2. Etching of organic materials. Etch rates, W, correspond to maximum values achieved. Most of the abbreviations employed are listed at the end of the book (see also caption to Table 5.1). The following additional contractions have been used: S etching of valleys; L etching of grooves; R etching of grooves; R etching of deep holes, including vias

Solid	Ambient	Laser λ [nm]	W [μm/s] (I [W/cm²]; ϕ [J/cm²]) d [μm]	Remarks	Ref.
Allyl-diglycol-carbonate ([$C_{12}H_{18}O_7$]$_n$)	Air	10.6 μm CO_2	(2.5E4)	L, P	[8.109]
Nitrocellulose (spun-on Si)		ArF KrF XeF	$W = 2$ μm/p (0.1) $d = 0.3$	P, λ	[8.110, 111]
PET	Air, vacuum	ArF 185 Hg lamp	$W = 0.12$ μm/p (0.15) (2.5E−3)	L, K, SEM, XPS	[8.112 – 115]
PET, PMMA, PI (Kapton[a])		ArF, XeCl	(≤0.8)	Stress measurements	[8.116]
PMMA	Air	ArF KrF	$W = 0.4$ μm/p (0.3) $W = 3.5$ μm/p (2)	P, λ, UV; model calc.	[8.97, 142]
PI	Air	ArF KrF XeCl	$W = 0.25$ μm/p (4) $W = 0.5$ μm/p (2) $W = 1.2$ μm/p (2)		
PET, PI	Air	ArF, KrF, XeCl	(≤0.3)	T	[8.117]
PMMA, PI, Kapton[a], Mylar	Air, He, vacuum	ArF	(≤1.4)	λ, UV, VIS, SEM	[8.15, 118]
PI	Air, He, vacuum	KrF	(≤2.7)		
PI, Kapton[a]	Air, He, vacuum	XeF	(≤3)		
PMMA	Air	ArF KrF	$W = 4$ μm/p (5) $W = 3$ μm/p (15)	P, λ, SEM; model calc.	[8.139]
PI	Air	ArF KrF	$W = 0.8$ μm/p (7) $W = 0.4$ μm/p (10)		

Solid	Ambient	Laser λ [nm]	W [µm/s] (I [W/cm^2]; ϕ [J/cm^2]) d [µm]	Remarks	Ref.
TNS2[b]	Air	ArF KrF	$W = 0.7$ µm/p (6) $W = 0.3$ µm/p (10)		[8.58]
PMMA		F$_2$, ArF	$d = 0.4$	SEM; projection patterning	[8.119 – 122]
PMMA	Air	ArF KrF	$W = 0.4$ µm/p (0.27) $W = 4.1$ µm/p (4.1)	S, P, λ, SEM; shape of holes	[8.123]
PMMA (PM20[c]) (spun-on Si)	Air	ArF	$W = 1$ µm/p (0.6) $d \leq 0.5$	SEM; projection patterning	[8.124]
PMMA + dye/PMMA		p-515 Ar + deep UV lamp	($\leq 1E7$)	S, P, SEM; laser lithography, bilayer resist structure	[8.125]
Polycarbonate	Air	ArF KrF XeCl	$W = 0.15$ µm/p $W = 0.60$ µm/p $W = 0.70$ µm/p	S, P, λ, K	[8.126]
PI	Air	KrF XeCl XeF	$W \leq 0.15$ µm/p (0.25) $W \leq 0.15$ µm/p (0.25) $W \leq 0.15$ µm/p (0.25)	P, λ, UV, VIS	[8.127]
PI Diazonaphtho-quinone (Novolak) (spun-on Si), PI (Kapton[a] foil)	Air	ArF	(0.02)	Projection patterning	
PI	Air	CO$_2$	(>2.1)	R, P, λ, IR	[8.140]

[a] Kapton is a Du Pont trademark polyimide.
[b] TNS2 is an IBM photoresist.
[c] PM20 is a cross-linked PMMA resist from Philips.

Fig.8.21. Scanning electron micrographs of polymers etched with ArF laser radiation through mechanical masks. Left: PMMA. Right: Riston (after ⌊8.114⌋)

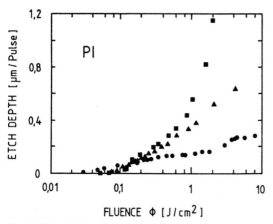

Fig.8.22. Etch depth per laser pulse in polyimide as a function of laser fluence for three different excimer laser wavelengths. Each data point was averaged over ten to several hundred laser pulses. •: 193 nm ArF; ▲: 248 nm KrF; ■: 308 nm XeCl (after ⌊8.142⌋)

The parameters that influence APD of polymers have been studied mainly with ArF, KrF and XeCl laser radiation. Taking the example of polyimide (PI), Fig.8.22 shows the etch depth per laser pulse as a function of fluence for three excimer laser wavelengths. A characteristic feature of APD is the occurrence of a threshold fluence of, typically, 0.01-0.1 J/cm^2, below which the laser radiation simply heats the polymer. This threshold decreases with decreasing laser wavelength. Above the threshold, etching commences but it

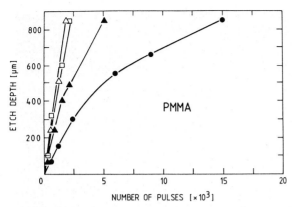

Fig.8.23. Depth of holes etched by ArF laser radiation in PMMA as a function of the number of laser pulses. $\triangle$: 0.475 J/cm^2; $\square$: 0.35 J/cm^2; $\blacktriangle$: 0.25 J/cm^2; $\bullet$: 0.15 J/cm^2 (after ⌊8.120⌋)

only lasts for as long as does the laser pulse. The etch depth per pulse, in general, increases with increasing laser wavelength.

The results obtained for the drilling of deep holes are shown in Fig.8.23 for the example of PMMA and ArF laser irradiation at various fluences ⌊8.120⌋. The width of such holes increases very slightly with increasing laser beam illumination time, i.e. with the number of laser pulses. It should be noted that for very deep holes, the profiles of holes become very complex and depend on the particular material and on the laser wavelength and fluence. For PMMA, holes with a simple conical shape (Fig.2.9b) have been obtained for ArF laser fluences above 0.350 - 0.475 J/cm^2. The etch depth increases linearly with the number of pulses, i.e. with time. For medium and low fluences a decrease in slope is observed for longer irradiation times (Fig.8.23). This drop-off in rate is most likely due to the diminution of the incident laser light intensity by the ablated material that is ejected out of the hole. This interpretation is supported by a simple estimate. The velocity of the ejected material was measured to be in the range of 10^5 - $7 \cdot 10^5$ cm/s, which is in reasonable agreement with the model calculations in ⌊8.128⌋. Under "steady-state" conditions, i.e. disregarding incubation effects observed during the first several laser pulses, one can estimate that the material will travel only about 10 to 100 μm during the laser pulse that lasts, typically, 16 ns. This means that for the etch depths shown in Fig.8.23, secondary photolysis of reaction products is important and becomes more efficient with increasing depth of holes and thereby with the number of laser pulses. The absence of the drop-off for the high laser fluences and depths of holes shown in Fig.8.23 may be related to the threshold for

197

ablation and/or to smaller fragments with smaller absorption cross sections, which may possibly occur at higher fluences. For the etching of deep holes, the real power dependence of the etch rate, however, is strongly masked by the aforementioned secondary absorption due to photofragments, by internal reflections within the hole etc. (see also Sect.2.1.2), and it cannot be revealed in a simple way from such measurements.

For the reasons outlined above, an analysis of *primary* ablated products in APD appears to be virtually impossible. The emission spectra in the visible and near ultraviolet spectral region originating from photofragments of various polymers (PI, PET, PMMA) exposed to ArF, KrF and XeF laser pulses have been investigated by various groups ⌊8.15,118,141⌋. While the intensity of the emission differed significantly for various ambient media (He, air, vacuum), the structure of the spectra was similar or the same for the different environments investigated. The spectra were also very similar for the various different polymers and laser wavelengths ⌊8.15,118⌋.

The patterning of nitrocellulose films spun onto Si wafers from a solution of nitrocellulose in amyl-acetate has been reported by DEUTSCH and GEIS ⌊8.110,111⌋. Figure 8.24 shows an optical micrograph of a grating pattern obtained with 193 nm ArF laser radiation and a mechanical mask contacted to the substrate. A resolution of better than 0.3 μm was achieved. For 248 nm KrF and 351 nm XeF laser radiation the surface morphology of the exposed region changed and numerous fibrous filaments occurred.

Projection patterning of organic layers has been demonstrated recently ⌊8.58,127⌋. With F_2 and ArF laser light a resolution of 0.2 to 0.4 μm has been achieved.

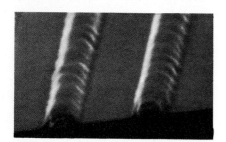 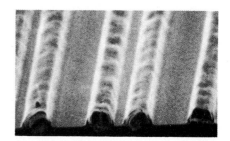

⊢ **2 μm** ⊣

Fig.8.24. Left: Optical micrograph of a grating pattern in nitrocellulose etched with ArF laser radiation and a 3.8 μm period chrome-on-quartz mask contacted to the substrate. Right: Grating pattern obtained when the mask is placed away from the substrate in order to obtain spatial period division effects (after ⌊8.111⌋)

9. Comparison of Processing Techniques, Applications of Laser Chemical Processing

Throughout this book so far we have mentioned quite a number of applications of LCP. Local and extended-area deposition, oxidation, nitridation, reduction, metallization, doping, compound formation and etching are needed in many areas of technology such as mechanics, electronics, integrated optics and chemical technology. In virtually all of these fields LCP offers new and unique processing possibilities, which are impossible with currently available technologies. On the other hand, there are many applications where LCP has to compete with standard and well-established techniques such as conventional chemical vapor deposition (CVD) ⌊9.1-4⌋; plasma processing, either deposition (PCVD) or etching (PE) ⌊9.3,4⌋; electron beam (EB) processing ⌊9.3-5⌋; ion beam processing, for example, reactive ion etching (RIE) ⌊9.3,4⌋; photochemical processing with lamps ⌊9.3,4⌋, etc.

This chapter is intended to summarize and discuss further the possibilities and limitations of LCP and to compare the different techniques with the corresponding standard technologies. Clearly, a partial replacement of standard technologies will not only depend on the quality of the processing technique, but also on its price, reliability, total throughput, etc. On the basis of the extremely rapid development of LCP within the past few years, however, an optimistic forecast would seem to be justified. Here, of course, future developments in laser technology will also play a decisive role.

The following discussion is divided into two parts: Two-dimensional *planar processing* and three-dimensional *nonplanar processing*. We define planar and nonplanar processing in the following way:

- If the substrate is planar and if the lateral dimensions of the processed feature are larger than, or comparable to, the axial (normal to the surface) dimensions, we define processing to be planar or two-dimensional. This includes large-area deposition and surface modification, but also most cases of compound formation, surface etching and direct writing on planar substrates.

199

- If the substrate is nonplanar and/or if the lateral dimensions of the deposited, transformed or etched material are small compared to the axial dimensions, we define processing to be nonplanar or three-dimensional. This includes the growth of rods or the etching of deep grooves, via holes, etc., but also all modes of processing on nonplanar substrates. Clearly, there are many intermediate cases which make classification somewhat arbitrary.

9.1 Planar Processing

As outlined in the foregoing chapters, LCP allows the deposition, modification, transformation or etching of thin extended films but also single-step direct writing of patterns. On the other hand, the conventional techniques mentioned above are all large-area processing techniques, with the exception of EB processing. With these latter techniques, patterns can be produced only in combination with mechanical masking or lithographic methods. Besides these differences from LCP, there are a number of other processing characteristics for each technique which are advantageous or disadvantageous in a particular application. These characteristics will now be discussed briefly.

9.1.1 Comparison of Processing Characteristics

For many applications, laser CVD is preferable to standard CVD simply because the processing temperature can be held lower and/or because any temperature increase can be confined to the area of processing. While standard CVD requires *uniform* substrate heating to, typically, 900-2000 K, LCVD can be performed at considerably lower temperatures and in some cases even at room temperature. Thus, LCVD allows deposition on temperature sensitive materials such as polymer foils, ceramics and compound semiconductors. These materials would melt, crack or dissociate at the temperatures required for many conventional CVD systems. Furthermore, even in cases where melting or cracking of the substrate surface is no problem, lower-temperature processing can be desirable because it avoids or reduces materials warpage, unwanted diffusion of species within the material or side reactions of the material's surface with the ambient medium. Clearly, diffusion and reaction rates depend exponentially on temperature (Chap.3). An example where elimination or reduction of diffusion is even a prior processing condition is deposition onto prefabricated Si surfaces for device applications. Here, lower processing temperatures reduce not only wafer warpage, but also dopant redistribution, defect generation and propagation, etc. These drawbacks can drastically reduce the yields in present-day chip fabrication and they will become even more severe with the ever-increasing demands being made on chips in terms of higher densities and, consequently, smaller feature sizes. Additionally, there are many materials that can be deposited with high quality at lower substrate temperatures, or which indeed may only be processed at lower temperatures. Examples are compound semiconductors and hydrogenated amorphous Si (Sect.5.3.2). For other materials, the purity, morphology and crystallinity of films may deteriorate with decreasing processing temperature and one has to strike a balance between the film quality and the temperature sensitivity of the substrate.

In plasma processing such as plasma CVD or plasma etching, a plasma is used to generate the reactive species. The processing temperatures for thin film deposition can thus be significantly lower than those used in standard CVD. Unfortunately, plasma techniques have a number of other inherent properties, which may be very disadvantageous or even make the technique inadequate for a particular application. Among these properties are heavy ion bombardment and VUV irradiation of the substrate, loading effects, and contamination of deposited or etched materials by impurities originating directly from the reactant and/or from carrier gases and/or from sputtering of the reactor chamber. Further problems very often arise from the difficulty of controlling processing parameters. The RF power, RF frequency, discharge geometry, electrode configuration, gas flow, total pressure, substrate temperature, etc. are all so interrelated that it is impossible to control and characterize effects due to a *single* parameter variation. Above all, a stable discharge can be maintained only for a very narrow range of operating parameters and this limits the versatility of the technique even further. The problems are similar in RIE. This latter technique involves ion-assisted chemical etching reactions between the material to be etched and reactive species, which again are generated by a plasma discharge.

Contrary to in PCVD, PE, RIE or EB processing, radiation damage, impurity sputtering, loading effects, etc. are absent or cause only minor problems in LCP. Additionally, LCP allows one to vary the laser power, laser wavelength, spatial location, gas flow, total pressure, substrate temperatures, etc. independently, i.e. without affecting any one of the other parameters. Furthermore, unlike ion or electron beams, laser radiation can be propagated through a great variety of media, or it can be made highly absorbable, e.g. by changing its wavelength. Additionally, lasers allow selective processing. For example, it is conceivable that alternating layers of elements or compounds can be deposited from admixtures of gases by simply changing the laser wavelength and thereby the decomposition of a particular species. Highly selective etching has already been realized for a number of systems. Examples are selective etching of Si on SiO_2 substrates $\lfloor 9.6\text{-}10 \rfloor$ or W on Si substrates $\lfloor 9.7\text{-}9 \rfloor$, where etching ratios (with regard to the substrate) of $10^2\text{-}10^3$ have been achieved.

As outlined in Sect.5.3, the deposition rates achieved in large-area LCVD are comparable to or even higher than those achieved in PCVD. The physical and chemical properties of laser-deposited films are already satisfactory in many respects (see e.g. Tables 5.2-4). Etch rates achieved in gas-phase LCP can already compete with those achieved in conventional dry-etching techniques. For example, with a 200-W excimer laser and a Cl_2 atmosphere it would be possible to remove Si from a 10 cm^2 area of a SiO_2 wafer or Si wafer with a rate of $2 \cdot 10^{-3}$ µm/s. This rate is similar to typical removal rates obtained in RF plasma etching. For W in a COF_2 atmosphere, ArF laser radiation with comparable intensity would yield an etch rate 3-4 times faster than PE $\lfloor 9.7\text{-}9 \rfloor$. For many other metals, PE is even less efficient than for W. The high etch rates achieved with laser light, in particular at normal incidence, are not only due to the photothermal or photochemical enhancement of the etching reaction itself, but also due to thermal or nonthermal desorption of nonvolatile reaction products, which are very often rate limiting in PE. Gas-or liquid-phase photochemical processing with lamps is often disadvantageous because the large volume irradiation results in the production of photofragments in regions which do not contribute to film growth or substrate etching. Consequently, reactants are not only lost, but they may also lead to unwanted reactions on reactor walls, etc. Furthermore, the reaction rates achieved with lamps are, in general, very low. For example, in the most efficient mercury-photosensitized reaction for the deposition of SiO_2, a deposition rate of $3 \cdot 10^{-4}$ µm/s has so far been achieved.

A different, unique application of lasers is single-step maskless localized processing either by spot-like illumination or by direct writing. The lateral dimensions of laser-processed patterns may reach down into the submicrometer range. The above-mentioned techniques currently used for pattern generation require, typically, 10-20 different dry- and wet-processing steps for a particular structure to be produced. Additionally, the repeated physical and chemical treatment influences the complete mechanical or electronical device in each cycle. For example, localized deposition requires several heating and cooling cycles and/or VUV irradiation or particle bombardment of the total device. Radiation damage and hazardous by-products in plasma etching may also cause problems. In liquid-phase etching, the exposure of the total device to an aggressive etching solution may cause serious damage as well. Problems associated with photoresist masks are sometimes also difficult to overcome. In other words, the conventional techniques will become problematic or even inadequate whenever physically or chemically sensitive materials, substrates or devices are to be processed. In many cases additional processing steps are even prohibited in order to avoid the destruction of a prefabricated device.

In LCP most or all of these problems are avoided. Physical treatment such as, for example, heat treatment or chemical treatment can be strongly localized, thereby leaving the material otherwise unaffected. These, and the other properties of LCP mentioned above, may become of fundamental importance in semiconductor VLSI or ULSI technology and a prior condition for fabrication of new multicomponent microdevices. Additionally, profiling for doping, etching, etc. achieved in LCP is often superior to that using conventional techniques.

It is evident that in all cases where standard techniques can be applied equally well or where the quality of a particular processing step can be tolerated, economical arguments will be decisive. Here, the most serious limitations of LCP at present are the total process rates and throughputs. While this applies in many cases of large-area laser processing it is even more pronounced in direct writing. As we have seen in the foregoing chapters, the local processing rates for deposition, etching, etc. can be extremely high, 100 μm/s or more. Nevertheless, the processed surface area per unit time is still quite small. Clearly, many of the conventional large-area techniques permit the fabrication of several devices simultaneously, despite being multiple step processes. In microelectronics, for example, thousands of devices that have been patterned by lithography on many separate chips can be produced in one series of operations. Therefore, in the foreseeable future, laser direct writing of *complete* complex structures may perhaps be interesting for the design of prototypes, but not for mass production.

To resolve this throughput problem one needs more powerful laser systems than those currently available, and further developments in optical projection and interference techniques that enable more efficient utilization of the laser energy. Therefore, LCP should be considered for the time being as a powerful *complementary* technique that can be used when standard techniques become inadequate. In such cases fabrication of tools, devices, wafers, etc. on a piece by piece basis becomes quite conceivable. However, even today there are already particular applications where LCP may substitute conventional processing techniques. Such applications essentially include all cases where small area complementations, modifications or repair of prefabricated devices or tools are necessary. In such cases the multiple step conventional techniques become very inefficient. Furthermore, localized deposition may be advantageous when recovery of precious materials such as rare metals from liquid or solid admixtures, for example, with photoresist materials produced as by-products in conventional processing, is expensive or altogether uneconomical.

9.1.2 Examples of Applications

In the following we shall look at a number of examples - either discussing them briefly or only mentioning them in passing - of potential industrial uses for large-area or localized LCP which stand a good chance of being realized in the near future, either because they are cheaper than conventional methods or because they represent unique and novel processing possibilities. Among those examples are: controlled-area hard coatings or protective coatings, solar cells, novel materials and controlled-grain-size materials with locally different physical and chemical properties, controlled-area material removal, one-step ohmic contacts, interconnects, localized ablative patterning, and lithography. Combination of the different possibilities allows mask and circuit repair, device restructuring and customization, circuit restructuring, design and fault correction, prototype electronic device fabrication, production of optical waveguides, etc.

a) Mainly Large-Area Processing

Insulating films such as SiO_2, Al_2O_3 or Si_3N_4 deposited by LCVD have potential applications as encapsulants, passivating layers and as dielectrics in electronic devices. Al_2O_3 is used in optical waveguides and as a coating for magnetic disks and disk heads; it is increasingly considered as an attractive substitute for SiO_2 in microelectronics, because it is a better barrier against dopants and Na ions, and also has a much greater resistance to radiation damage. Some of these applications benefit from a low temperature controlled-area high deposition rate process.

Silicides have great importance for protection against corrosion and in integrated circuit technology. Metal silicides can provide ohmic or rectifying contacts, interconnects, barriers to interdiffusion, etc. [9.55].

Conventional vacuum furnace production of metal silicides often results in interpenetration depths in excess of those allowable in IC technology.

Laser doping makes it possible to produce very shallow dopant profiles with concentrations up to the solid solubility limit (Table 6.2). The processing times are very rapid, of the order of microseconds. Direct laser doping is a single-step process and seems to be cheaper when compared with the conventional techniques of ion implantation and subsequent oven annealing or rapid thermal annealing (RTA). This has been used to fabricate pn junctions, for example for solar cell production. For Si solar cells with areas of up to several square centimeters, AM1 (sunlight illumination) efficiencies of 10-13% have been achieved [9.11-15]. Gallium arsenide solar cells produced by laser doping had an efficiency of nearly 11% [9.16-18].

The production of extended thin films of a-Si:H by CO_2 laser-induced deposition seems to be very promising. Such films can be used for low cost solar cells, thin film transistors, etc. [9.19-22]. Here, an optimistic forecast seems appropriate not only on the basis of the high quality of films already achieved, but also on the basis of prices and throughputs. CO_2 lasers have a very high efficiency (typically around 20%), are commercially available at powers into the multikilowatt range and are also among the most reliable and durable laser sources.

Laser lithography is receiving increasing attention. F_2 and ArF excimer lasers (Sect.4.2) can be used for single-step ablation of organic photoresists (Sect.8.4). Since exposure of current resists is limited to wavelengths ⩾ 300 nm, short wavelength excimer laser radiation will improve the achievable resolution of patterns by at least a factor of 2 (Sect.3.5). Furthermore, with excimer laser light a steeper image profile can be achieved, because the modulation transfer function of a lithographic system increases with decreasing wavelength [9.23]. Additionally, excimer lasers make it possible to use *new* resists that can be designed to have higher contrast and improved physical and chemical properties than those presently used. The high average power of excimer lasers also makes it possible to increase the throughput with regard to conventional photolithography. This is not only due to the multiple steps needed in the standard technique but also to the low efficiency of lamps used for exposure of relatively insensitive UV resists. For example, with a commercially available contact printer, exposure times for a 10 cm (4 in.) wafer were, typically, 2-3 min with ArF laser radiation, compared to 20 min with an UV lamp in the same system [9.24]. The sensitivity of resists acceptable for lamp source irradiation also limits the variety of materials and their optimization for other processing steps.

Recent developments have been reviewed by JAIN et al. ⌊9.25,26⌋. Both contact and proximity printing have been used to demonstrate submicrometer resolution. Projection lithography has been performed by using a commercial scanning projection printer. Full-wafer exposure with 1 μm resolution and excellent image quality has been illustrated [9.27]. With F_2 laser light a resolution better than 0.2 μm has been achieved ⌊9.28,29⌋. Patterning of inorganic resists has been demonstrated as well ⌊9.28,30-32,56⌋.

b) Mainly Localized Processing

One-step ohmic contacts of metals, metal silicides, poly-Si, etc. can be produced at high speed by addressed spot-like illumination. The electrical properties of such contacts have been characterized in greater detail for Au-InP ⌊9.33⌋, Pt-GaAs ⌊9.33⌋ and for directly metallized oxides [9.34,35,54]. Ohmic contact formation by direct surface doping has been investigated for a large number of systems (Table 6.2).

Direct writing of conducting lines can be used for rapid production of short interconnects, for circuit design, restructuring and customization, for mask and circuit repair, etc. ⌊9.36-43⌋. The possibilities of the technique have been demonstrated for a number of devices. $Cd(CH_3)_2$ photodecomposition has been used to directly write Cd gate electrodes on Si MOSFETs [9.38,39]. DEUTSCH ⌊9.44⌋ used a combination of Mo-Al gate metallization, which served as a mask for ArF laser-induced B doping of n-Si substrates using BCl_3 as doping gas. The process allows single-step self-aligned doping of MOSFETs. Laser microchemical restructuring has been demonstrated by direct writing of poly-Si links on a ring oscillator ⌊9.38,39⌋ and on gate array structures ⌊9.40,41⌋.

Photodeposition has also been successfully used for single-step repair of transparent defects on hard-surface photomasks for optical lithography ⌊9.45⌋. Repair of pinholes and breaks by photodeposition can be achieved with an accuracy of better than 1 μm.

Gratings produced by deposition ⌊9.46,47⌋ or etching [9.48-50] reactions within the field of two interfering laser beams can be used for many electro-optical devices ⌊9.51⌋ and for integrated optical circuits in general.

Localized or selective materials etching can be used in microfabrication for patterning, restructuring, customization, trimming, etc. High spatial resolution and/or selectivity is often needed to avoid perturbation or destruction of nearby structures, devices or substrate films. An example of the latter case is the trimming of polycrystalline Si or metal conductors over dielectric oxide or nitride films.

Optical waveguides have been produced in $LiNbO_3$ by photodeposition of Ti from $TiCl_4$ and subsequent Ti indiffusion ⌊9.52⌋.

9.2 Nonplanar Processing

The technique of LCP offers unique possibilities for nonplanar three-dimensional processing. For example, it makes it possible to produce uniform protective coatings on mechanical workpieces of complex geometrical structure, on glass fibers for optical communication, etc. One can also conceive of many applications in chemical technology. For example, for catalytic purposes, production of metal coatings on cheap substrates, such as glass wool, could become of importance. Here, high porosity sponge-like deposition would be possible by an appropriate control of the processing parameters. Also, LCP makes it possible to write virtually any structure directly onto complicated three-dimensional devices. Here, patterning by mechanical masking or lithographic techniques together with large-area processing is extremely difficult and in most cases even impossible. Furthermore, LCP makes it possible to *selectively* increase the thickness of protective coatings (produced by standard techniques) on the edges or sides of tools or devices where mechanical or chemical requirements are particularly severe. Millimeter-high wall-like structures of various different materials can be directly written with lateral dimensions of only a few micrometers. Rods, of essentially any length can be grown, even as single crystals, without any crucible and in an otherwise completely cold atmosphere. This technique should make it possible not only to grow materials with higher purity, but also to grow new materials, quite possibly under extreme physical conditions, for example at very high temperatures, pressures, or within electric or magnetic fields. Here, applications can only be speculated on. The growth of rods also makes it possible to investigate very rapidly the reaction kinetics and properties of systems that are relevant in standard CVD (Sect.5.2.2). Laser-induced deposition also allows three-dimensional growth and packaging by means of pattern projection, interference techniques, etc. Clearly, deep grooves or holes can be filled in with a great variety of materials.

Laser-induced chemical etching could be used for efficient lower temperature high resolution trimming of oscillators and resistors, etc. Removal of organic materials from tools and devices, such as stripping of organic insulators from wires, is easily performed by APD, as described in Sect.8.4. Another application would be the etching of deep grooves or holes, including vias. Examples are via holes for mechanical devices such as nozzles for high pressure jet tooling, ink-printing, etc. Through-wafer via conductors may be of importance in novel IC technologies. They could be produced simply by etching and consecutive filling by just changing the reactive surrounding. For example, etching of a via hole through a commercial Si wafer would take only about 25 s and still retain an aspect ratio of about 8:1 ⌊9.10⌋. Consecutive refilling with a metal would take a similar time ⌊9.53⌋. Laser chemical processing could significantly reduce dimensions and spacings of such grooves or via holes.

List of Abbreviations and Symbols

The following list covers the main abbreviations and notation used throughout the text and within the tables. The meaning of single symbols, however, may be somewhat different in each case. For example, within the text, P refers to the effective laser power, while within the tables it means that the dependence on laser power has been investigated.

a	aperture
AcAc	$\lfloor CH_3COCHCOCH_3 \rfloor^-$ = 2,4-pentane dionato anion = acetylacetonate anion
AES	Auger electron spectroscopy
AL	adsorbed layers
APD	ablative photodecomposition
a-X	amorphous material of composition X
c	speed of light
c_p	specific heat at constant pressure
CARS	coherent antistokes Raman scattering
CP	chemical properties
CVD	chemical vapor deposition
c-X	single-crystalline material of composition X
D	molecular diffusivity $\lfloor cm^2/s \rfloor$
D_t	thermal diffusivity $\lfloor cm^2/s \rfloor$
d	lateral width of laser-processed features $\lfloor \mu m \rfloor$
E	energy; $k_B T(T = 273.15\ K) = 2.3538 \cdot 10^{-2}$ eV
ΔE	activation energy $\lfloor kcal/mole \rfloor$
e	electron charge
EB	electron beam
EBCVD	electron beam-induced chemical vapor deposition
EELS	electron energy loss spectroscopy
EF	extended thin film formation or etching, large-area processing
EM	enhancement of reaction rate
EMF	electromotoric force

ESCA	electron spectroscopy for chemical analysis
ESR	electron spin resonance
eV	electron Volt; 1 eV = $8.0655 \cdot 10^3$ cm^{-1} = $1.1604 \cdot 10^4$ K; 1 eV/particle = 23.06 Kcal/mole
F	electric field $\lfloor V \rfloor$
f	focal length $\lfloor \mu m, cm \rfloor$
FH	fourth harmonic
FWHM	full width at half maximum
h	Planck's constant; height, thickness or depth of laser-processed features $\lfloor A, \mu m \rfloor$
h_L	thickness of free standing film or layer (e.g. evaporated) on substrate
HFAcAc	$\lfloor CF_3COCHCOCF_3 \rfloor^-$ = 1,1,1,5,5,5 hexafluoro-2,4-pentane dionato anion = hexafluoroacetylacetonate anion
HS	heated substrate (temperature in $\lfloor K \rfloor$)
HV	high vacuum (p > 10^{-7} mbar)
I	laser intensity $\lfloor W/cm^2 \rfloor$
IR	infrared radiation; absorption or emission spectroscopy of samples or species in the infrared spectral range
ITO	indium tin oxide
K	kinetics
k	kinetic constant $\lfloor$ see e.g. (3.5)$\rfloor$
k_B	Boltzmann's constant
k_L	wave vector of laser beam
Kapton	polyimide (Du Pont)
L	direct writing of stripes or grooves; length of laser focus
LCP	laser-induced chemical processing
LCVD	laser-induced CVD
LEE	laser-enhanced electrochemical etching
LEED	low energy electron diffraction
LEP	laser-enhanced electrochemical plating
LIF	laser-induced fluorescence
LPCVD	laser-enhanced PCVD
LPE	laser-enhanced plasma etching
LPPC	laser-pulsed plasma chemistry
ML	laser used in multiline operation
MOCVD	metal-organic CVD
MPD	multiphoton dissociation
MPI	multiphoton ionization

MS	mass spectroscopy
n	refractive index
n_c	dopant concentration
OMA	optical multichannel analyzer
P	laser power $\lfloor W \rfloor$
p	pressure $\lfloor mbar \rfloor$
p_i	partial pressure of species i $\lfloor mbar \rfloor$
p_0	vapor pressure $\lfloor mbar \rfloor$
p_{tot}	total gas pressure
PCVD	plasma CVD
PE	plasma etching
PET	polyethylene terephthalate
PI	polyimide
PL	photoluminescence
PLZT	lanthanum doped PZT
PMMA	polymethyl methacrylate
PM20	cross-linked PMMA resist (Philips)
PP	general physical properties
pps	pulses per second
PVDF	polyvinylidene fluoride
PXE	same as PZT
Pyrex	borosilicate glass
PZT	lead titanate zirconate $PbTi_{1-x}Zr_xO_3$
p-X	polycrystalline material of composition X
QCM	quartz crystal microbalance
R	optical reflection coefficient; gas constant $\lfloor J/K\ mole \rfloor$
R_D	optical reflection coefficient of deposited material
RA	Raman spectroscopy; diagnostics using Raman spectroscopy
RBS	Rutherford backscattering spectroscopy
RF	radio frequency
RHEED	reflection high-energy electron diffraction
RIE	reactive ion etching
Riston	negative photoresist consisting of PMMA and other acrylates (Du Pont)
rms	root mean square
RTA	rapid thermal annealing
S	deposition or etching of spots
SEM	scanning electron microscopy
SERS	surface enhanced Raman scattering
SH	second harmonic

SI	semiinsulating
SIMS	secondary ion mass spectroscopy
T	temperature $\lfloor K \rfloor$
T_c	center temperature $\lfloor K \rfloor$
T_m	melting temperature $\lfloor K \rfloor$
T_s	substrate temperature $\lfloor K \rfloor$
T_t	threshold temperature $\lfloor K \rfloor$
ΔT	temperature rise $\lfloor K \rfloor$
t_d	laser beam dwell time $\lfloor s \rfloor$
t_i	laser beam illumination time $\lfloor s \rfloor$
t_n	nucleation time $\lfloor s \rfloor$
TEM	transmission electron microscopy
TH	third harmonic
TiBAl	$Al(C_4H_9)_3$
TOF	time-of-flight
UHV	ultrahigh-vacuum ($p < 10^{-7}$ mbar)
ULSI	ultra-large-scale integrated systems
UV	ultraviolet radiation; absorption or emission spectroscopy of samples or species in the ultraviolet spectral range
v	lateral growth velocity $\lfloor \mu m/s \rfloor$
v_g	velocity of gas flow
v_s	scanning velocity of laser beam or substrate $\lfloor \mu m/s \rfloor$
VIS	visible radiation; absorption or emission spectroscopy of samples or species in the visible spectral range
VLSI	very-large-scale intergrated systems
VUV	vacuum UV
W	reaction rate (deposition, etching etc.) $\lfloor \mu m/s, \text{Å/pulse} \rfloor$
w	$w_0/2^{1/2}$ radius of laser focus (1/e intensity) $\lfloor \mu m \rfloor$
w_0	radius of laser focus ($1/e^2$ intensity) $\lfloor \mu m \rfloor$
X	X-ray
XEY	$X \cdot 10^Y$
XPS	X-ray photoemission spectroscopy
z_R	Rayleigh length $\lfloor \mu m \rfloor$
α	optical absorption coefficient $\lfloor cm^{-1} \rfloor$
β	exchange coefficient
γ	reaction order
δ	delta function
ε	emissivity

ε_d dielectric constant; dielectric properties

η dissociation yield

κ thermal conductivity, deposit (κ_D), thin layer (κ_L), substrate (κ_S) $\lfloor W/mK \rfloor$

λ wavelength $\lfloor nm, \mu m \rfloor$

ν frequency $\lfloor s^{-1} \rfloor$

ξ overpotential

ρ electrical resistivity, deposit (ρ_D), bulk (ρ_B) $\lfloor \Omega cm \rfloor$

ρ_m mass density $\lfloor g/cm^3 \rfloor$

σ electrical conductivity $\lfloor \Omega^{-1} cm^{-1} \rfloor$; cross section $\lfloor cm^2 \rfloor$

σ_a absorption cross section $\lfloor cm^2 \rfloor$

σ_d dissociation cross section $\lfloor cm^2 \rfloor$

τ laser pulse time; relaxation time $\lfloor s \rfloor$

Φ electrical potential

ϕ laser fluence $\lfloor J/cm^2 \rfloor$

Ψ wave function

ω angular frequency $\lfloor s^{-1} \rfloor$

$\perp$ laser light incident normal to substrate

$\parallel$ laser light incident parallel to substrate

References

Chapter 1

1.1 A.E. Siegman: *An Introduction to Lasers and Masers* (McGraw-Hill, New York 1971)
1.2 F.T. Arecchi, E.O. Schulz-Dubois (eds.): *Laser Handbook 1,2* (North-Holland, Amsterdam 1972)
1.3 M.L. Stitch (ed.): *Laser Handbook 3* (North-Holland, Amsterdam 1979)
1.4 H.Haken: *Licht und Materie II, Laser* (Bibliographisches Institut, Mannheim 1981)
1.5 A. Yariv: *Quantum Electronics* (Wiley, New York 1967)
1.6 D. Bäuerle (ed.): *Laser Processing and Diagnostics,* Springer Ser. Chem. Phys., Vol. 39 (Springer, Berlin, Heidelberg 1984)
1.7 J.F. Ready: *Industrial Applications of Lasers* (Academic, New York 1978)
1.8 J.F. Ready: Proc. IEEE 70, 533 (1982)
1.9 G. Herziger, E.W. Kreutz: in ⌊1.6⌋, p. 90
1.10 E. Beyer, P. Loosen, R. Poprawe, G. Herziger: Laser und Optoelektronik 3, 274 (1985)
1.11 D. Schuöcker (ed.): *Industrial Applications of High Power Lasers,* SPIE, Vol. 455 (1983); Laser und Optoelektronik 1, 55 (1986)
1.12 H. Koebner: *Industrial Applications of Lasers* (Wiley, New York 1984)
1.13 W.M. Steen (ed.): *Advances in Cutting Processes* (Welding Institute, Cambridge, U.K. 1986)
1.14 M. Bass (ed.): *Laser Material Processing* (North-Holland, Amsterdam 1983)
1.15 D.W. Draper: J. Metals, June, 24 (1982)
1.16 M. v. Allmen (ed.): *Amorphous Metals and Nonequilibrium Processing* (Physique, Les Ulis 1984)
1.17 J. Götzlich, H. Ryssel: in ⌊1.6⌋, p. 40
1.18 S.D. Ferries, H.J. Leamy, J.M. Poate (eds.): *Laser-Solid Interactions and Laser Processing* (AIP, New York 1979)
1.19 C.W. White, P.S. Peercy (eds.): *Laser and Electron Beam Processing of Materials* (Academic, New York 1980)
1.20 J.F. Gibbons, L.D. Hess, T.W. Sigmon (eds.): *Laser and Electron Beam Solid Interactions and Materials Processing* (North-Holland, New York 1981)
1.21 J.M. Poate, J.W. Mayer (eds.): *Laser Annealing of Semiconductors* (Academic, New York 1982)
1.22 J. Narayan, W.L. Brown, R.A. Lemons (eds.): *Laser-Solid Interactions and Transient Thermal Processing of Materials* (North-Holland, New York 1983)
1.23 A.G. Cullis: Rep. Prog. Phys. 48, 1155 (1985)
1.24 B.R. Appleton, G.K. Celler: *Laser and Electron-Beam Interactions with Solids* (North-Holland, New York 1982)
1.25 K. Hennig (ed.): *Energy Pulse Modification of Semiconductors and Related Materials* (Akademie der Wissenschaften, Dresden 1984)

1.26 V.T. Nguyen, A.G. Cullis: *Energy Beam-Solid Interactions and Transient Thermal Processing* (Physique, Les Ulis 1985)
1.27 I. Ursu, A.M. Prokhorov (eds.): *Trends in Quantum Electronics,* Abstracts (European Physical Society, Bucharest 1985)
1.28 R.J. von Gutfeld, J.E. Greene, H. Schlossberg (eds.): *Beam Induced Chemical Processes,* Extended Abstracts (Materials Research Society, Boston 1985)
1.29 D. Bäuerle, K.L. Kompa, L.D. Laude (eds.): *Laser Processing and Diagnostics II* (Physique, Les Ulis 1986)
1.30 D. Bäuerle, L.D. Laude, M. Wautelet (eds.): *Interfaces under Laser Irradiation,* Nato ASI Series (M. Nijhoff, Dordrecht 1987)
1.31 R.M. Osgood, S.R.J. Brueck, H.R. Schlossberg (eds.): *Laser Diagnostics and Photochemical Processing for Semiconductor Devices* (North-Holland, New York 1983)
1.32 D. Bäuerle: in *Surface Studies with Lasers,* ed. by F.R. Aussenegg, A. Leitner, M.E. Lippitsch, Springer Ser. Chem. Phys., Vol. 33 (Springer, Berlin, Heidelberg 1983) p. 178
1.33 A.W. Johnson, D.J. Ehrlich, H.R. Schlossberg (eds.): *Laser Controlled Chemical Processing of Surfaces* (North-Holland, New York 1984)
1.34 T.J. Chuang: *Laser-Induced Chemical Etching of Solids* (to be published by Springer, Berlin, Heidelberg)
1.35 D. Bäuerle: Laser und Optoelektronik $\underline{1}$, 29 (1985)
1.36 S. Metev (ed.): *Laser Assisted Modification and Synthesis of Materials,* Proc. Int. Winter School, Gyuletchitza, Bulgaria 1985
1.37 D.J. Ehrlich, J.Y. Tsao: J. Vac. Sci. Technol. B $\underline{1}$, 969 (1983)
1.38 D.Bäuerle: in $\lfloor 1.26 \rfloor$, p. 103
1.39 S.D. Allen (ed.): *Laser Assisted Deposition, Etching and Doping,* Proc. SPIE Conf. $\underline{459}$ (1984)
1.40 D. Bäuerle: in $\lfloor \overline{1.6} \rfloor$, p. 166
1.41 R.J. von Gutfeld: Laser Appl. $\underline{5}$, 1 (1984)
1.42 F.A. Houle, T.F. Deutsch, R.M. Osgood (eds.): *Laser Chemical Processing of Semiconductor Devices,* Extended Abstracts (Materials Research Society, Boston 1984)
1.43 Y. Rytz-Froidevaux, R.P. Salathé, H.H. Gilgen: Appl. Phys. A $\underline{37}$, 121 (1985)
1.44 R. Solanki, C.A. Moore, G.J. Collins: Solid State Technol. June, 220 (1985)
1.45 A. Ben-Shaul, Y. Haas, K.L. Kompa, R.D. Levine: *Lasers and Chemical Change,* Springer Ser. Chem. Phys., Vol. 10 (Springer, Berlin, Heidelberg 1981)
1.46 K.L. Kompa, S.D. Smith (eds.): *Laser-Induced Processes in Molecules,* Springer Ser. Chem. Phys., Vol. 6 (Springer, Berlin, Heidelberg 1979)
1.47 A.H. Zewail (ed.): *Advances in Laser Chemistry,* Springer Ser. Chem. Phys., Vol. 3 (Springer, Berlin, Heidelberg 1978)
1.48 J.I. Steinfeld (ed.): *Laser-Induced Chemical Processes* (Plenum, New York 1981)
1.49 V.S. Letokhov: *Nonlinear Laser Chemistry, Multiple Photon Excitation,* Springer Ser. Chem. Phys., Vol. 22, (Springer, Berlin, Heidelberg 1983)
1.50 J.H. Clark, K.M. Leary, T.R. Loree, L.B. Harding: in $\lfloor 1.47 \rfloor$, p.74
1.51 M.W. Berns: Laser Focus $\underline{19}$, 66 (1983)
1.52 J. Hecht: Lasers and Applications, May, 79 (1986)
1.53 D. Muller: Lasers and Applications, May, 85 (1986)
1.54 R.J. Lane, J.J. Wynne: Lasers and Applications, November,59 (1984)
1.55 R.J. Lane, R. Linsker, J.J. Wynne, A. Torres, R.G. Geronemous: Arch. Dermatol. (1985)

Chapter 2

2.1 D. Bäuerle: in *Laser Processing and Diagnostics,* ed. by D. Bäuerle, Springer Ser. Chem. Phys., Vol. 39 (Springer, Berlin, Heidelberg 1984) p. 166
2.2 K. Piglmayer, J. Doppelbauer, D. Bäuerle: in *Laser Controlled Chemical Processing of Surfaces,* ed. by A.W. Johnson, D.J. Ehrlich, H.R. Schlossberg (North-Holland, New York 1984) p. 47
2.3 F. Petzoldt, K. Piglmayer, W. Kräuter, D. Bäuerle: Appl. Phys. A 35, 155 (1984)
2.4 K. Piglmayer, D. Bäuerle: in *Laser Processing and Diagnostics II,* ed. by D. Bäuerle, K.L. Kompa, L.D. Laude (Physique, Les Ulis 1986)
2.5 K. Piglmayer: Ph. D. Thesis, Linz (1986)
2.6 M. Eyett, D. Bäuerle: unpublished
2.7 M. Lax: J. Appl. Phys. 48, 3919 (1977)
2.8 M. Lax: Appl. Phys. Lett. 33, 786 (1978)
2.9 H.E. Cline, T.P. Anthony: J. Appl. Phys. 48, 3895 (1977)
2.10 Y.I. Nissim, A. Lietola, R.B. Gold, J.F. Gibbons: J. Appl. Phys. 51, 274 (1980)
2.11 J.E. Moody, R.H. Hendel: J. Appl. Phys. 53, 4364 (1982)
2.12 F. Stern: J. Appl. Phys. 44, 4204 (1973)
2.13 M. Sparks: J. Appl. Phys. 47, 837 (1976)
2.14 F. Ferrieu, G. Auvert: J. Appl. Phys. 54, 2646 (1983)
2.15 R.F. Wood, G.E. Giles: Phys. Rev. B 23, 2923 (1981)
2.16 J.H. Batteh: J. Appl. Phys. 53, 7537 (1982)
2.17 D.J. Sanders: Appl. Optics 23, 30 (1984)
2.18 S.A. Kokorowski, G.L. Olsen, L.D. Hess: in ⌊2.35⌋, p. 139
2.19 F.G. Allen: J. Appl. Phys. 28, 1510 (1957)
2.20 Y.J. Van der Meulen, N.C. Hein: J. Opt. Soc. Am. 64, 804 (1964)
2.21 M.O. Lambert, J.M. Koebel, P. Siffert: J. Appl. Phys. 52, 4975 (1981)
2.22 G.E. Jellison, F.A. Modine: Phys. Rev. B 27, 7466 (1983)
2.23 D.L. Kwong, D.M. Kim: J. Appl. Phys. 54, 366 (1983)
2.24 D.L. Kwong, D.M. Kim: IEEE J. QE-18, 224 (1982)
2.25 D.M. Kim, D.L. Kwong, R.R. Shah, D.L. Crosthwait: J. Appl. Phys. 52, 4995 (1981)
2.26 D.M. Kim, D.L. Kwong, R.R. Shah, D.L. Crosthwait: in ⌊2.33⌋, p. 83
2.27 I.D. Calder, R. Sue: J. Appl. Phys. 53, 7545 (1982)
2.28 J.P. Colinge, F. Van de Wiele: J. Appl. Phys. 52, 4769 (1981)
2.29 H.S. Carslaw, T.C. Jaeger: *Conduction of Heat in Solids* (Oxford University Press, New York 1959)
2.30 See e.g., J.M. Ziman: *Principles of the Theory of Solids* (Cambridge University Press, London 1972)
2.31 C.Y. Ho, R.W. Powell, P.E. Liley: J. Phys. Chem. 3, Suppl. 1, 588 (1974)
2.32 J.F. Osmundsen, C.C. Abele, J.G. Eden: in ⌊2.2⌋, p. 259
2.33 S.D. Ferries, H.J. Leamy, J.M. Poate (eds.): *Laser-Solid Interactions and Laser Processing* (AIP, New York 1979)
2.34 C.W. White, P.S. Peercy (eds.): *Laser and Electron Beam Processing of Materials* (Academic, New York 1980)
2.35 J.F. Gibbons, L.D. Hess, T.W. Sigmon (eds.): *Laser and Electron Beam Solid Interactions and Materials Processing* (North-Holland, New York 1981)
2.36 J.M. Poate, J.W. Mayer (eds.): *Laser Annealing of Semiconductors* (Academic, New York 1982)
2.37 J. Narayan, W.L. Brown, R.A. Lemons (eds.): *Laser-Solid Interactions and Transient Thermal Processing of Materials* (North-Holland, ·New York 1983)
2.38 A.G. Cullis: Rep. Prog. Phys. 48, 1155 (1985)

2.39 B.R. Appleton, G.K. Celler: *Laser and Electron-Beam Interactions with Solids* (North-Holland, New York 1982)
2.40 D. Bäuerle, L.D. Laude, M. Wautelet (eds.): *Interfaces under Laser Irradiation,* Nato ASI Series (M. Nijhoff, Dordrecht 1987)
2.41 G.E. Jellison, F.A. Modine: Appl. Phys. Lett. *41*, 180 (1982)
2.42 I.W. Boyd, T.D. Binnie, J.I.B. Wilson, M.J. Colles: J. Appl. Phys. *55*, 3061 (1984)
2.43 S. Unamuno, M. Toulemonde, P. Siffert: in ⌊2.1⌋, p. 35
2.44 W.L. Brown: in ⌊2.34⌋, p. 20; H.M. van Driel: in ⌊2.40⌋
2.45 D. Marsal: *Die Numerische Lösung Partieller Differentialgleichungen* (Bibliographisches Institut, Mannheim 1976)
2.46 A.J. Davies: *The Finite Element Method* (Clarendon, Oxford 1980)
2.47 R.T. Kivaisi, L. Stensland: Appl. Phys. A *27*, 233 (1982)
2.48 A. Ben-Shaul, Y. Haas, K.L. Kompa, R.D. Levine: *Lasers and Chemical Change,* Springer Ser. Chem. Phys., Vol. 10 (Springer, Berlin, Heidelberg 1981)
2.49 K.L. Kompa, S.D. Smith (eds.): *Laser-Induced Processes in Molecules,* Springer Ser. Chem. Phys., Vol. 6 (Springer, Berlin, Heidelberg 1979)
2.50 A.H. Zewail (ed.): *Advances in Laser Chemistry,* Springer Ser. Chem. Phys., Vol. 3 (Springer, Berlin, Heidelberg 1978)
2.51 J.I. Steinfeld (ed.): *Laser-Induced Chemical Processes* (Plenum, New York 1981)
2.52 V.S. Letokhov: *Nonlinear Laser Chemistry, Multiple Photon Excitation,* Springer Ser. Chem. Phys., Vol. 22, (Springer, Berlin, Heidelberg 1983)
2.53 G. Herzberg: *Molecular Spectra and Molecular Structure, I. Spectra of Diatomic Molecules* (Van Nostrand Reinhold, New York 1950)
2.54 G. Herzberg: *Molecular Spectra and Molecular Structure, II. Infrared and Raman Spectra of Polyatomic Molecules* (Van Nostrand Reinhold, New York 1945)
2.55 G. Herzberg: *Molecular Spectra and Molecular Structure, III. Electronic Spectra and Electronic Structure of Polyatomic Molecules* (Van Nostrand Reinhold, New York 1966)
2.56 R. Bersohn: IEEE J. QE-*16*, 1208 (1980)
2.57 J.G. Calvert, J.N. Pitts: *Photochemistry* (Wiley, New York 1966)
2.58 H. Walther, K.W. Rothe (eds.): *Laser Spectroscopy IV,* Springer Ser. Opt. Sci., Vol. 21 (Springer, Berlin, Heidelberg 1979)
2.59 J.L. Hall, J.L. Carlsten (eds.): *Laser Spectroscopy III,* Springer Ser. Opt. Sci., Vol. 7 (Springer, Berlin, Heidelberg 1977)
2.60 J.H. Eberly, P. Lambropoulos (eds.): *Multiphoton Processes* (Wiley, New York 1978)
2.61 C.D. Cantrell (ed.): *Multiple Photon Excitation and Dissociation of Polyatomic Molecules,* Topics Curr. Phys., Vol. 35 (Springer, Berlin, Heidelberg 1986)
2.62 D.O. Cowan, R.L. Drisko: *Elements of Organic Photochemistry* (Plenum, New York 1976)
2.63 D.C. Hanna, M.A. Yuratich, D. Cotter: *Nonlinear Optics of Free Atoms and Molecules,* Springer Ser. Opt. Sci., Vol. 17 (Springer, Berlin, Heidelberg 1979)
2.64 S.J.W. Price: in *Decomposition of Inorganic and Organometallic Compounds,* ed. by C.H. Bamford, C.F.H. Tipper, Chemical Kinetics, Vol. 4 (Elsevier, Amsterdam 1972) p. 197
2.65 T. Motooka, S. Gorbatkin, D. Lubben, J.E. Greene: J. Appl. Phys. *58*, 4397 (1985)
2.66 J.H. Clark, R.G. Anderson: Appl. Phys. Lett. *32*, 46 (1978)
2.67 J.T. Yardley, B. Gitlin, G. Nathanson, A. Rosan: J. Chem. Phys. *74*, 370 (1981)
2.68 Y. Rytz-Froidevaux, R.P. Salathé, H.H. Gilgen: in *Laser Diagnostics and Photochemical Processing for Semiconductor Devices,* ed. by R.M. Osgood, S.R.J. Brueck, H.R. Schlossberg (North-Holland, New York 1983) p. 29

2.69   D.B. Geohegan, A.W. McCown, J.G. Eden: in ⌊2.2⌋, p. 93
2.70   H.H. Gilgen, T. Cacouris, P.S. Shaw, R.R. Krchnavek, R.M. Osgood: Appl.
       Phys. A, in press
2.71   M. Hirose, S. Yokoyama, Y. Yamakage: J. Vac. Sci. Technol. B 3, 1445
       (1985)
2.72   D.K. Flynn, J.I. Steinfeld, D.S. Sethi: J. Appl. Phys. 59, 3914 (1986)
2.73   H. Schröder: in *Laser Assisted Modification and Synthesis of Materials,*
       Proc. Int. Winter School, Gyuletchitza, Bulgaria 1985, ed. by S. Metev,
       p. 120
2.74   Q. Mingxin, R. Monot, H.v.d. Bergh: Scientia Sinica-A 1, 85 (1985)
2.75   H.H. Gilgen, C.J. Chen, R. Krchnavek, R.M. Osgood: in ⌈2.1⌋, p. 225
2.76   M. Tamir, U. Halavee, R.D. Levine: Chem. Phys. Lett. 25, 38 (1974)
2.77   C. Jonah, P. Chandra, R. Bersohn: J. Chem. Phys. 55, 1903 (1971)
2.78   M.A. Duncan, T.G. Dietz, R.E. Smalley: Chem. Phys. 44, 415 (1979)
2.79   T.A. Seder, S.P. Church, A.J. Ouderkirk, E. Weitz: J. Am. Chem. Soc.
       107, 1432 (1985)
2.80   W.H. Breckenridge, N. Sinai: J. Chem. Phys. 85, 3557 (1981)
2.81   Z. Karny, R. Naaman, R.N. Zare: Chem. Phys. Lett. 59, 33 (1978)
2.82   D.P. Gerrity, L.J. Rothberg, V. Vaida: Chem. Phys. Lett. 74, 1 (1980)
2.83   G.J. Fisanick, A. Gedanken, T.S. Eichelberger, N.A. Kuebler,
       M.B. Robin: J. Chem. Phys. 75, 5215 (1981)
2.84   T.R. Fletcher, R.N. Rosenfeld: J. Am. Chem. Soc. 105, 6358 (1983)
2.85   E.E. Marinero, C.R. Jones: J. Chem. Phys. 82, 1608 (1985)
2.86   H. Schröder, I. Gianinoni, D. Masci, K.L. Kompa in ⌊2.1⌋, p. 257
2.87   M.C. Heaven, M.A.A. Clyne: J. Chem. Soc., Faraday. Trans. 2, 78,
       1339 (1982)
2.88   M.A.A. Clyne, I.S. McDermick: J. Chem. Soc., Faraday Trans. 2, 75, 1677
       (1979)
2.89   S. Leone: J. Phys. Chem. 85, 3844 (1981)
2.90   G.L. Loper, M.D. Tabat: Appl. Phys. Lett. 46, 654 (1985)
2.91   G.L. Loper, M.D. Tabat: SPIE 459, 121 (1984)
2.92   G.L. Loper, M.D. Tabat: J. Appl. Phys. 58, 3649 (1985)
2.93   D.A. Armstrong, J.L. Holmes: in ⌊2.64⌋, p. 143
2.94   T.F. Deutsch, D.D. Rathman: Appl. Phys. Lett. 45, 623 (1984)
2.95   T.F. Deutsch: in ⌊2.1⌋, p. 239
2.96   G.G.A. Perkins, E.R. Austin, F.W. Lampe: J. Am. Chem. Soc. 101, 1109
       (1979)
2.97   M.P. Irion, K.L. Kompa: Appl. Phys. B 27, 183 (1982)
2.98   N.R. Greiner: J. Chem. Phys. 47, 4373 (1967)
2.99   K.F. Preston, R.J. Cvetanovic: in ⌊2.64⌋, p. 47
2.100 M. Zelikoff, L.M. Aschenbrand: J. Chem. Phys. 52, 1680 (1974)
2.101 H.S. Johnston, R. Craham: Can. J. Chem. 52, 1680 (1974)
2.102 K.F. Preston, R.F. Barr: J. Chem. Phys. 54, 3347 (1971)
2.103 J. Zavelovich, M. Rothschild, W. Gornik, C.K. Rhodes: J. Chem. Phys.
       74, 6787 (1981)
2.104 V.M. Donnelly, A.P. Boronanski, J.R. McDonald: Chem. Phys. 43,
       271 (1979)
2.105 H. Niki, G.J. Mains: J. Phys. Chem. 68, 304 (1964)
2.106 Y. Rousseau, G.J. Mains: J. Phys. Chem. 70, 3158 (1966)
2.107 S.J.C. Irvine, J.B. Mullin, J. Tunnicliffe: in ⌊2.1⌋, p. 234
2.108 S.J.C. Irvine, J.B. Mullin, J. Tunnicliffe: J. Cryst. Growth 68, 188
       (1984)
2.109 N. Bloembergen, E. Yablanovitch: Phys. Today, May, 23 (1978)
2.110 V.S. Letokhov, A.A. Makarov: Sov. Phys.-Usp. 24, 366 (1981)
2.111 P.J. Robinson, K.A. Holbrook: *Unimolecular Reactions* (Wiley, New York
       1972)
2.112 R.V. Ambartsumyan, V.S. Letokhov: in *Chemical and Biochemical
       Applications of Lasers,* Vol. 3, ed. by C.B. Moore (Academic, New York
       1977) p. 167

2.113 D.M. Cox: Optics Commun. 24, 336 (1978)
2.114 S.D. Smith, W.E. Schmid, F.M.G. Tablas, K.L. Kompa: in ⌊2.49⌋, p. 121
2.115 M. Meunier, J.H. Flint, D. Adler, J.S. Haggerty: in ⌊2.2⌋, p. 397
2.116 T.F. Deutsch: J. Chem. Phys. 70, 1187 (1979)
2.117 P.A. Longeway, F.W. Lampe: J. Am. Chem. Soc. 103, 6813 (1981)
2.118 M. Hanabusa, A. Namiki, K. Yoshihara: Appl. Phys. Lett. 35, 626 (1979)
2.119 M. Hanabusa, H. Kikuchi, T. Iwanaga, K. Sugai: in ⌊2.1⌋, p. 197
2.120 M. Hanabusa, H. Kikuchi, T. Iwanaga, K. Sugai: in ⌊2.2⌋, p. 21
2.121 R.V. Ambartsumyan, Y.A. Gorokhov, V.S. Letokhov, G.N. Markarov,
      A.A. Puretskii: Sov. Phys.-JETP 44, 231 (1976)
2.122 I. Burk, P. Houston, D.G. Sutton, J.I. Steinfeld: J. Chem. Phys.
      53, 3632 (1970)
2.123 H.S. Kwok, E. Yablonovitch: Phys. Rev. Lett. 41, 745 (1978)
2.124 W. Fuss, T.P. Cotter: Appl. Phys. 12, 265 (1977)
2.125 V.N. Bagratashvili, I.N. Kuyazev, V.S. Letokhov, V.V. Lobko:
      Opt. Commun. 18, 525 (1976)
2.126 P.A. Schulz, A.S. Sudbo, E.R. Grant, Y.R. Shen, Y.T. Lee:
      J. Chem. Phys. 72, 4985 (1980)
2.127 E.R. Grant, M.J. Coggiola, Y.T. Lee, P.A. Schulz, A. Sudbo,
      Y.R. Shen: Chem. Phys. Lett. 52, 595 (1977)
2.128 P.A. Schulz, A.S. Sudbo, D.J. Krajnovich, H.S. Kwok, Y.R. Shen,
      Y.T. Lee: Annu. Rev. Phys. Chem. 30, 379 (1979)
2.129 T.J. Chuang: IBM J. Res. Dev. 26, 145 (1982)
2.130 F.A. Houle, T.J. Chuang: J. Vac. Sci. Technol. 20, 790 (1982)
2.131 T.J. Chuang: in ⌊2.2⌋, p. 185
2.132 T.J. Chuang: in ⌊2.68⌋, p. 45
2.133 T.J. Chuang: J. Chem. Phys. 74, 1453 (1981)
2.134 V.N. Bagratashvili, V.S. Doljikov, V.S. Letokhov, E.A. Ryabov: Zh.
      Tekh. Fiz., Pis'ma Red. 4, 1181 (1978)
2.135 I.P. Herman, J.B. Marling: Chem. Phys. Lett. 64, 75 (1979)
2.136 H. Schröder, S. Metev, W. Robers, B. Rager: in ⌊2.4⌋

Chapter 3

3.1 J.G. Dash: *Films on Solid Surfaces* (Academic, New York 1975)
3.2 D.M. Young, A.D. Crowell: *Physical Adsorption of Gases* (Butterworths,
     London 1962)
3.3 G. Benedek, U. Valbusa (eds.): *Dynamics of Gas-Surface Interaction*,
     Springer Ser. Chem. Phys., Vol. 21 (Springer, Berlin, Heidelberg 1982);
     G. Benedek: in ⌊3.38⌋
3.4 F.R. Aussenegg, A. Leitner, M.E. Lippitsch (eds.): *Surface Studies with
     Lasers*, Springer Ser. Chem. Phys., Vol. 33 (Springer, Berlin, Heidelberg
     1983)
3.5 D. Bäuerle (ed.): *Laser Processing and Diagnostics*, Springer Ser. Chem.
     Phys., Vol. 39 (Springer, Berlin, Heidelberg 1984)
3.6 A. Terenin: Adv. Catal. 15, 227 (1964)
3.7 A. Nitzan, L.E. Brus: J. Chem. Phys. 75, 2205 (1981)
3.8 J. Garsten, A. Nitzan: J. Chem. Phys. 73, 3023 (1980)
3.9 C.J. Chen, R.M. Osgood: Phys. Rev. Lett. 50, 1705 (1983)
3.10 C.J. Chen, R.M. Osgood: Appl. Phys. A 31, 171 (1983)
3.11 T.F. George: J. Phys. Chem. 86, 10 (1982); T.F. George, D. Jelski,
     Xi-Yi Huang, A.C. Beri: in ⌊3.38⌋
3.12 J. Heidberg, H. Stein, A. Nestmann, E. Hoefs, I. Hussla: in *Laser-Solid
     Interactions and Laser Processing*, ed. by S.D. Ferries, H.J. Leamy,
     J.M. Poate (AIP, New York 1979) p. 49
3.13 J. Heidberg, H. Stein, A. Nestmann, E. Hoefs, I. Hussla: Surf. Sci. 126,
     183 (1983)

3.14 T.J. Chuang: J. Vac. Sci. Technol. B 3, 1408 (1985)
3.15 T.J. Chuang, H. Seki, I. Hussla: Surf. Sci. 158, 525 (1985)
3.16 D. Bäuerle: in ⌊3.5⌋, p. 166
3.17 T.H. Wood, J.C. White, B.A. Thacker: Appl. Phys. Lett. 42, 408 (1983)
3.18 T.H. Wood, J.C. White, B.A. Thacker: in *Laser Diagnostics and Photoche-mical Processing for Semiconductor Devices,* ed. by R.M. Osgood, S.R.J. Brueck, H.R. Schlossberg (North-Holland, New York 1983) p. 35
3.19 U.M. Titulaer, D. Bäuerle: to be published
3.20 D.J. Ehrlich, J.Y. Tsao: J. Vac. Sci. Technol. B 1, 969 (1983)
3.21 P.M. George, J.L. Beauchamp: Thin Solid Films 67, L 25 (1980);
     A.E. Stevens, C.S. Feigerle, W.C. Lineberger: J. Am. Chem. Soc. 104 (1982)
3.22 J.Y. Tsao, R.A. Becker, D.J. Ehrlich, F.J. Leonberger: Appl. Phys. Lett. 42, 559 (1983)
3.23 H. Schröder: in *Laser Assisted Modification and Synthesis of Materials,* Proc. Int. Winter School, Gyuletchitza, Bulgaria 1985, ed. by S. Metev, p. 120
3.24 H.F. Winters, J.W. Coburn, T.J. Chuang: J. Vac. Sci. Technol. A 1, 1157 (1983)
3.25 H.F. Winters, J.W. Coburn, T.J. Chuang: J. Vac. Sci. Technol. B 1, 469 (1983)
3.26 R. Gauthier, C. Guittard: Phys. Stat. Sol. A 38, 477 (1976)
3.27 R.J. von Gutfeld: in ⌊3.5⌋, p. 323
3.28 R.J. von Gutfeld: Denki Kagaku 52, 452 (1984)
3.29 See, e.g., J.O.M. Bockris, A.K.N. Reddy: *Modern Electrochemistry I,II* (Plenum, New York 1977)
3.30 W.E. Langlois: private communication
3.31 R.J. von Gutfeld, M.H. Gelchinski, L.T. Romankiw, D.R. Vigliotti: Appl. Phys. Lett. 43, 876 (1983)
3.32 J.C. Puippe, R.E. Acosta, R.J.von Gutfeld: J. Electrochem. Soc. 128, 2539 (1981)
3.33 J.C. Puippe, R.E. Acosta, R.J.von Gutfeld: Oberfläche-Surf. 22, 294 (1981)
3.34 R.J.von Gutfeld, R.E. Acosta, L.T. Romankiw: IBM J. Res. Dev. 26, 136 (1982)
3.35 H. Gerischer: Electroanal. Chem. Interfacial Electrochem. 58, 263 (1975)
3.36 F.W. Ostermayer, P.A. Kohl, R.M. Lum: J. Appl. Phys. 58, 4390 (1985)
3.37 P.G. Shewmon: *Diffusion in Solids* (McGraw-Hill, New York 1963)
3.38 H.P. Bonzel: in *Interfaces under Laser Irradiation,* ed. by D. Bäuerle, L.D. Laude, M. Wautelet, Nato ASI Series (M. Nijhoff, Dordrecht 1987)
3.39 L.C. Kimerling, J.L. Benton: in *Laser and Electron Beam Processing of Materials,* ed. by C.W. White, P.S. Peercy (Academic, New York 1980) p. 385
3.40 P.C. Townsend: in *Sputtering in Particle Bombardment II,* ed. by R. Behrish (Springer, Heidelberg, Berlin 1983) p. 147
3.41 J.E. Rothenberg, R. Kelly: Nucl. Inst. Meth. Phys. Res. B 1, 291 (1984)
3.42 R. Kelly, J.J. Cuomo, P.A. Leary, J.E. Rothenberg, B.E. Braren, C.F. Aliotta: Nucl. Inst. Meth. Phys. Res. B 9, 329 (1985)
3.43 N. Itoh: in ⌊3.38⌋
3.44 R. Srinivasan: in ⌊3.38⌋
3.45 E. Sutcliffe, R. Srinivasan: J. Appl. Phys. (1986)
3.46 T. Keyes, R.H. Clarke, J.M. Isner: J. Phys. Chem. 89, 4194 (1986)
3.47 R. Srinivasan: in ⌊3.5⌋, p. 343
3.48 B.J. Garrison, R. Srinivasan: J. Appl. Phys. 57, 2909 (1985)
3.49 H.H.G. Jellinek, R. Srinivasan: J. Phys. Chem. 88, 3048 (1984)
3.50 R. Srinivasan, P.E. Dyer, B. Braren: to be published
3.51 G. Koren, J.T.C. Yeh: Appl. Phys. Lett. 44, 1112 (1984)
3.52 G. Koren, J.T.C. Yeh: J. Appl. Phys. 56, 2120 (1984)

3.53 K. Piglmayer, D. Bäuerle: unpublished
3.54 See, e.g., L.I. Maissel, R. Glang: *Handbook of Thin Film Technology*
(McGraw-Hill, New York 1970)
3.55 J.H. Brannon, J.R. Lankard, A.I. Baise, F. Burns, J. Kaufman: J. Appl.
Phys. $\underline{58}$, 2036 (1936)

Chapter 4

4.1 D. Bäuerle: Laser und Optoelektronik $\underline{1}$, 29 (1985)
4.2 D. Bäuerle (ed.): *Laser Processing and Diagnostics,* Springer Ser. Chem.
Phys., Vol. 39 (Springer, Berlin, Heidelberg 1984)
4.3 R.J. von Gutfeld: in $\lfloor 4.2 \rfloor$, p. 323
4.4 R.J. von Gutfeld: Denki Kagaku $\underline{52}$, 452 (1984)
4.5 A.E. Siegman: *An Introduction to Lasers and Masers* (McGraw-Hill, New
York 1971)
4.6 W. Demtröder: *Laser Spectroscopy,* Springer Ser. Chem. Phys., Vol. 5
(Springer, Berlin, Heidelberg 1981)
4.7 K. Thyagarajan, A.K. Ghatak: *Lasers, Theory and Applications* (Plenum,
New York 1981)
4.8 A.G. Cullis, H.C. Webber, P. Bailey: J. Phys. E $\underline{12}$, 688 (1979)
4.9 E. Söllner, E. Benes, A. Biedermann, D. Hammer: Vacuum $\underline{27}$, 367 (1977)
4.10 E. Benes: J. Appl. Phys. $\underline{56}$, 608 (1984)
4.11 S. Hertl, L. Wimmer, E. Benes: in $\lfloor 4.2 \rfloor$, p. 464
4.12 G. Kreuer, L.A. Moraga: Rev. Sci. Instrum. $\underline{56}$, 1467 (1985)
4.13 T.J. Chuang: IBM J. Res. Dev. $\underline{26}$, 145 (1982)
4.14 F.A. Houle, T.J. Chuang: J. Vac. Sci. Technol. $\underline{20}$, 790 (1982)
4.15 T.J. Chuang: J. Chem. Phys. $\underline{74}$, 1453 (1981)
4.16 R.L. Melcher: in $\lfloor 4.2 \rfloor$, p. 418
4.17 G. Leyendecker, H. Noll, D. Bäuerle, P. Geittner, H. Lydtin:
J. Electrochem. Soc. $\underline{130}$, 157 (1983)
4.18 J. Doppelbauer, D. Bäuerle: in *Laser Processing and Diagnostics II,* ed.
by D. Bäuerle, K.L. Kompa, L.D. Laude (Physique, Les Ulis 1986)
4.19 W. Kräuter, D. Bäuerle, F. Fimberger: Appl. Phys. A $\underline{31}$, 13 (1983)
4.20 B. Braren, R. Srinivasan: J. Vac. Sci. Technol. B $\underline{3}$, 913 (1985)
4.21 P.E. Dyer, R. Srinivasan: Appl. Phys. Lett. $\underline{48}$, 445 (1986)
4.22 C. Karner, A. Mandel, F. Träger: Appl. Phys. A $\underline{38}$, 19 (1985)
4.23 B.C. Larson, C.W. White, T.S. Noggle, D. Mills: Phys. Rev. Lett. $\underline{48}$
337 (1982)
4.24 H.W. Lo, A. Compaan: Phys. Rev. Lett. $\underline{44}$, 1604 (1980)
4.25 L.A. Lompre, J.M. Liu, H. Kurz, N. Bloembergen: Appl. Phys. Lett. $\underline{43}$,
1163 (1983)
4.26 A. Moritani, C. Hamaguchi: Appl. Phys. Lett. $\underline{46}$, 746 (1985)
4.27 M.A. Bösch, R.A. Lemons: Phys. Rev. Lett. $\underline{47}$, 1151 (1981)
4.28 T.O. Sedgwick: Mater. Res. Soc. Symp. Proc. $\underline{1}$, 147 (1981)
4.29 T.O. Sedgwick: Appl. Phys. Lett. $\underline{39}$, 254 (1981)
4.30 R.P. Salathé, H.H. Gilgen: in $\lfloor 4.2 \rfloor$, p. 425
4.31 M. Kemmler, G. Wartmann, D. von der Linde: Appl. Phys. Lett. $\underline{45}$, 159
(1984)
4.32 J. Doppelbauer, D. Bäuerle: to be published
4.33 G. Alestig, G. Holmen: J. Appl. Phys. $\underline{56}$, 2161 (1984)
4.34 D. Bäuerle, P. Irsigler, G. Leyendecker, H. Noll, D. Wagner: Appl.
Phys. Lett. $\underline{40}$, 819 (1982)

Chapter 5

5.1 R.K. Montgomery, T.D. Mantei: Appl. Phys. Lett. $\underline{48}$, 493 (1986)
5.2 A. Auerbach: J. Electrochem. Soc. $\underline{132}$, 130 (1985)

5.3 A. Auerbach: J. Electrochem. Soc. 132, 1437 (1985)
5.4 G.S. Higashi, L.J. Rothberg: J. Vac. Sci. Technol. B 3, 1460 (1985)
5.5 R. Solanki, W.H. Ritchie, G.J. Collins: Appl. Phys. Lett. 43,
     454 (1983)
5.6 P.K. Boyer, C.A. Moore, R. Solanki, W.K. Ritchie, G.A. Roche,
     G.J. Collins: in *Laser Diagnostics and Photochemical Processing for
     Semiconductor Devices,* ed. by R.M. Osgood, S.R.J. Brueck, H.R. Schloss-
     berg (North-Holland, New York 1983) p. 119
5.7 A.R. Calloway, T.A. Galantowicz, W.R. Fenner: J. Vac. Sci. Technol.
     A 1, 534 (1983)
5.8 D.J. Ehrlich, R.M. Osgood,Jr., T.F. Deutsch: J. Vac. Sci. Technol.
     21, 23 (1982)
5.9 D.J. Ehrlich, R.M. Osgood,Jr., T.F. Deutsch: Appl. Phys. Lett.
     38, 946 (1981)
5.10 D.J.Ehrlich, R.M. Osgood, Jr.: Thin Solid Films 90, 287 (1982)
5.11 T.F. Deutsch, D.J. Ehrlich, R.M. Osgood, Jr.: Appl. Phys. Lett. 35,
     175 (1979)
5.12 R.M. Osgood,Jr., D.J. Ehrlich: Opt. Lett. 7, 385 (1982)
5.13 D.J. Ehrlich, R.M. Osgood, T.F. Deutsch: IEEE J. QE-16, 1233 (1980)
5.14 D.J. Ehrlich, R.M. Osgood, Jr. : Chem. Phys. Lett. 79, 381 (1981)
5.15 S.R.J. Brueck, D.J. Ehrlich: Phys. Rev. Lett. 48, 1678 (1982)
5.16 J.Y. Tsao, D.J. Ehrlich: in *Laser Controlled Chemical Processing of
     Surfaces,* ed. by A.W. Johnson, D.J. Ehrlich, H.R. Schlossberg
     (North-Holland, New York 1984) p. 115
5.17 J.Y. Tsao, D.J. Ehrlich: Appl. Phys. Lett. 46, 198 (1985)
5.18 J.E. Bourée, Y.I. Nissim, J. Flicstein, C. Licoppe, R. Druilhe:
     in *Energy Beam-Solid Interactions and Transient Thermal Processing,*
     ed. by V.T. Nguyen, A.G. Cullis (Physique, Les Ulis 1985) p. 119
5.19 Y. Rytz-Froidevaux, R.P. Salathé, H.H. Gilgen: Phys. Lett. 84 A, 216
     (1981)
5.20 Y. Rytz-Froidevaux, R.P. Salathé: in *Laser Assisted Deposition,
     Etching and Doping,* ed. by S.D. Allen, Proc. SPIE Conf. 459, 55 (1984)
5.21 J.Y. Tsao, D.J. Ehrlich: Appl. Phys. Lett. 45, 617 (1984)
5.22 D.B. Geohegan, A.W. McCown, J.G. Eden: in ⌊5.16⌋, p. 93
5.23 J.Y. Tsao, D.J. Ehrlich: J. Chem. Phys. 81, 4620 (1984)
5.24 J.Y. Tsao, D.J. Ehrlich: in ⌊5.20⌋, p. 2
5.25 T.F. Deutsch, D.J. Silversmith, R.W. Mountain: in ⌊5.6⌋, p. 129
5.26 T.F. Deutsch: in *Laser Processing and Diagnostics,* ed. by D.
     Bäuerle, Springer Ser. Chem. Phys., Vol. 39 (Springer, Berlin,
     Heidelberg 1984) p. 239
5.27 T.F. Deutsch, D.J. Silversmith, R.W. Mountain: in ⌊5.16⌋, p. 67
5.28 K. Emery, P.K. Boyer, L.R. Thompson, R. Solanki, H. Zarnani,
     G.J. Collins: in ⌊5.20⌋, p. 9
5.29 Z. Yu, G.J. Collins, R. Solanki: in *Beam Induced Chemical Processes,*
     Extended Abstracts, ed. by R.J. von Gutfeld, J.E. Greene, H. Schloss-
     berg, (Materials Research Society, Boston 1985) p. 93
5.30 T.H. Baum, C.R. Jones: Appl. Phys. Lett. 47, 538 (1985)
5.31 T.H. Baum, C.R. Jones: in ⌊5.29⌋, p. 3
5.32 R.J. von Gutfeld: in ⌊5.26⌋, p. 323
5.33 R.J. von Gutfeld: Denki Kagaku 52, 452 (1984)
5.34 R.J. von Gutfeld, M.H. Gelchinski, L.T. Romankiw, D.R. Vigliotti:
     Appl. Phys. Lett 43, 876 (1983)
5.35 R.J. von Gutfeld, M.H. Gelchinski, D.R. Vigliotti, L.T. Romankiw: in
     ⌊5.16⌋, p. 325
5.36 M.H. Gelchinski, R.J. von Gutfeld, L.T. Romankiw, D.R. Vigliotti:
     J. Electrochem. Soc. 132, 2575 (1985)
5.37 S. Tamir, J. Zahavi: J. Vac. Sci. Technol. A 3, 2312 (1985)
5.38 J.C. Puippe, R.E. Acosta, R.J. von Gutfeld: J. Electrochem. Soc. 128,
     2539 (1981)

5.39 J.C. Puippe, R.E. Acosta, R.J. von Gutfeld: Oberfläche-Surf. 22, 294 (1981)
5.40 R.J. von Gutfeld, R.E. Acosta, L.T. Romankiw: IBM J. Res. Dev. 26, 136 (1982)
5.41 R.J. von Gutfeld, E.E. Tynan, R.L. Melcher, S.E. Blum: Appl. Phys. Lett. 35, 651 (1979)
5.42 R.F. Karlicek, V.M. Donnelly, G.J. Collins: J. Appl. Phys. 53, 1084 (1982)
5.43 R.H. Micheels, A. D. Darrow, R.D. Rauh: Appl. Phys. Lett. 39, 418 (1981)
5.44 A.K. Al-Sufi, H.J. Eichler, J.Salk, H.J. Riedel: J. Appl. Phys 54, 3629 (1983)
5.45 G.F. Fisanick, M.E. Gross, J.B. Hopkins, M.D. Fennell, K.J. Schnoes, A. Katzir: J. Appl. Phys. 57, 1139 (1985)
5.46 G.F. Fisanick, M.E. Gross, J.B. Hopkins, M.D. Fennell, K.J. Schnoes, A. Katzir: Appl. Phys. Lett. 46, 1184 (1985)
5.47 M.D. Fennell, G.J. Fisanick, D.K. Atwood: in ⌊5.29⌋, p. 39
5.48 D. Bäuerle: in ⌊5.26⌋, p. 166
5.49 G. Leyendecker, D. Bäuerle, P. Geittner, H. Lydtin: Appl. Phys. Lett. 39, 921 (1981)
5.50 D. Bäuerle: in ⌊5.6⌋, p. 19
5.51 G. Leyendecker, H. Noll, D. Bäuerle, P. Geittner, H. Lydtin: J. Electrochem. Soc. 130, 157 (1983)
5.52 D. Bäuerle, G. Leyendecker, P. Geittner, H. Lydtin: Appl. Phys. B 28, 267 (1982)
5.53 J. Doppelbauer, D. Bäuerle: in *Laser Processing and Diagnostics II*, ed. by D. Bäuerle, K.L. Kompa, L.D. Laude (Physique, Les Ulis 1986)
5.54 G. Leyendecker, H. Noll, D. Bäuerle: unpublished
5.55 L.S. Nelson, N.L. Richardson: Mater. Res. Bull 7, 971 (1972)
5.56 W. Pompe, H.J. Scheibe, G. Kessler, A. Richter, H.J. Weiss: in *Trends in Quantum Electronics,* Abstracts, ed. by I. Ursu, A.M. Prokhorov (European Physical Society, Bucharest 1985) p. 178
5.57 D.J. Ehrlich, J.Y. Tsao: in ⌊5.26⌋, p. 386
5.58 T.H. Wood, J.C. White, B.A. Thacker: Appl. Phys. Lett. 42, 408 (1983)
5.59 T.H. Wood, J.C. White, B.A. Thacker: in ⌊5.6⌋, p. 35
5.60 C.J. Chen, R.M. Osgood: Phys. Rev. Lett. 50, 1705 (1983)
5.61 C.J. Chen, R.M. Osgood: Appl. Phys. A 31, 171 (1983)
5.62 H.H. Gilgen, D. Podlesnik, C.J. Chen, R.M. Osgood, Jr.: in ⌈5.16⌋, p. 139
5.63 H.H. Gilgen, C.J. Chen, R. Krchnavek, R.M. Osgood: in ⌊5.26⌋, p. 225
5.64 M.W. Jones, L.J. Rigby, D. Ryan: Nature (London) 8, 177 (1966)
5.65 Y. Rytz-Froidevaux, R.P. Salathé, H.H. Gilgen, H.P. Weber: Appl. Phys. A 27, 133 (1982)
5.66 T.L. Rose, D.H. Longendorfer, R.D. Rauh: Appl. Phys. Lett. 42, 193 (1983)
5.67 W.M. Steen: in Int. Conf. on Advances in Surface Coating (London, 1978) p. 175
5.68 R. Solanki, P.K. Boyer, J.E. Mahan, G.J. Collins: Appl. Phys. Lett. 38, 572 (1981)
5.69 R. Solanki, P.K. Boyer, G.J. Collins: Appl. Phys. Lett. 41, 1048 (1982)
5.70 H. Yokoyama, F. Uesugi, S. Kishida, K. Washio: Appl. Phys. A 37, 25 (1985)
5.71 D.J. Ehrlich, R.M. Osgood, Jr., T.F. Deutsch: J. Electrochem. Soc. 128, 2039 (1981)
5.72 T.M. Mayer, G.J. Fisanick, T.S. Eichelberger IV: J. Appl. Phys. 53, 8462 (1982)
5.73 C. Arnone, M. Rothschild, J.G. Black, D.J. Ehrlich: Appl. Phys. Lett. 48, 1018 (1986)

5.74 F.A. Houle, C.R. Jones, T. Baum, C. Pico, C.A. Kovac: Appl. Phys. Lett. 46, 204 (1985)

5.75 C.R. Moylan, T.H. Baum, C.R. Jones: Appl. Phys. A 40, 1 (1986)

5.76 D. Braichotte, K. Ernst, R. Monot, J.M. Philippoz, M. Qiu, H.v.d. Bergh: Frühjahrstagung der Schweiz 58, 879 (1985); Q. Mingxin, R. Monot, H.v.d. Bergh: Scientia Sinica A 27, 531 (1984)

5.77 C.R. Jones, F.A. Houle, C.A. Kovac, T.H. Baum: Appl. Phys. Lett. 46, 97 (1985)

5.78 R.J. von Gutfeld, D.R. Vigliotti: Appl. Phys. Lett. 46, 1003 (1985)

5.79 R. B. Gerassimov, S.M. Metev, S.K. Savtchenko, G.A. Kotov, V.P. Veiko: Appl. Phys. B 28, 266 (1982)

5.80 P.J. Love, R.T. Loda, P.R. LaRoe, A.K. Green, Victor Rehn: in ⌊5.16⌋, p. 101

5.81 N. Bottka, P.J. Walsh, R.Z. Dalbey: J. Appl. Phys. 54, 1104 (1983)

5.82 J.S. Foord, R.B. Jackman: Chem. Phys. Lett. 112, 190 (1984)

5.83 R.B. Jackman, J.S. Foord, A.E. Adams, M.L. Lloyd: J. Appl. Phys. 59, 2031 (1986); R.B. Jackman, J.S. Foord: in ⌊5.53⌋

5.84 S.D. Allen, A.B. Trigubo: J. Appl. Phys. 54, 1641 (1983)

5.85 S.D. Allen, A.B. Trigubo, R.Y. Jan: in ⌊5.6⌋, p. 207

5.86 Y. Rytz-Froidevaux, R.P. Salathé, H.H. Gilgen: in ⌊5.6⌋, p. 29

5.87 M.R. Aylett, J. Haigh: in ⌊5.6⌋, p. 177

5.88 H. Beneking: in ⌊5.26⌋, p. 188

5.89 W. Roth, H. Kräutle, A. Krings, H. Beneking: in ⌊5.6⌋, p. 193

5.90 Y. Aoyagi, S. Masuda, S. Namba: Appl. Phys. Lett. 47, 95 (1985)

5.91 S.M. Bedair, J.K. Whisnant, N.H. Karam, M.A. Tischler, T. Katsuyama: Appl. Phys. Lett. 48, 174 (1986)

5.92 D. Bäuerle, S. Szikora, G. Constant, F. Maury: unpublished

5.93 R.W. Andreatta, C.C. Abele, J.F. Osmundsen, J.G. Eden, D. Lubben, J.E. Greene: Appl Phys. Lett. 40, 183 (1982)

5.94 R.W. Andreatta, D. Lubben, J.G. Eden, J.E. Greene: J. Vac. Sci. Technol. 20, 740 (1982)

5.95 J.G. Eden, J.E. Greene, J.F. Osmundsen, D. Lubben, C.C. Abele, S. Gorbatkin, H.D. Desai: in ⌊5.6⌋, p. 185

5.96 J.F. Osmundsen, C.C. Abele, J.G. Eden: in ⌊5.16⌋, p. 259

5.97 S.J.C. Irvine, J.B. Mullin, J. Tunnicliffe: in ⌊5.26⌋, p. 234

5.98 S.J.C. Irvine, J.B. Mullin, J. Tunnicliffe: J. Cryst. Growth 68, 188 (1984)

5.99 V.M. Donnelly, D. Braren, A. Appelbaum, M. Geva: J. Appl. Phys. 58, 2022 (1985)

5.100 V.M. Donnelly, M. Geva, J. Long, R.F. Karlicek: Appl. Phys. Lett. 44, 951 (1984)

5.101 V.M. Donnelly, M. Geva, J. Long, R.F. Karlicek: in ⌊5.16⌋, p. 73

5.102 A. Kitai, G.J. Wolga: in ⌊5.6⌋, p. 141

5.103 H.H. Gilgen, T. Cacouris, P.S. Shaw, R.R. Krchnavek, R.M. Osgood: Appl. Phys. A (1986)

5.104 D. Bäuerle: in Surface Studies with Lasers, ed. by F.R. Aussenegg, A. Leitner, M.E. Lippitsch, Springer Ser. Chem. Phys., Vol. 33 (Springer, Berlin, Heidelberg 1983) p. 178

5.105 W. Kräuter, D. Bäuerle, F. Fimberger: Appl. Phys. A 31, 13 (1983)

5.106 W. Kräuter, D. Bäuerle: unpublished

5.107 I.P. Herman, R.A. Hyde, B.M. McWilliams, A.H. Weisberg, L.L. Wood: in ⌊5.6⌋, p. 9

5.108 F. Petzoldt , K. Piglmayer, W. Kräuter, D. Bäuerle: Appl. Phys. A 35, 155 (1984)

5.109 S.D. Allen, M. Bass: J. Vac. Sci. Technol. 16, 431 (1979)

5.110 S.D. Allen: in Laser Applications in Materials Processing, SPIE 198, 49 (1979)

5.111 S.D. Allen, R.Y. Jan, S.M. Mazuk, K.J. Shin, S.D. Vernon: in ⌊5.16⌋, p. 1

5.112 S.D. Allen: J. Appl. Phys. 52, 6501 (1981)
5.113 *Metal Finishing 85* (Metals and Plastics Publications, Hackensack, NJ 1985)
5.114 L.J. Rigby: Trans. Faraday Soc. 65, 2421 (1969)
5.115 M.E. Gross: in ⌊5.29⌋, p. 21
5.116 H. Schröder: in *Laser Assisted Modification and Synthesis of Materials*, Proc. Int. Winter School, Gyuletchitza, Bulgaria 1985, ed. by S. Metev, p. 120
5.117 H. Schröder, I. Gianinoni, D. Masci, K.L. Kompa in ⌊5.26⌋, p. 257
5.118 H. Schröder, K.L. Kompa, D. Masci, I. Gianinoni: Appl. Phys. A. 38, 227 (1985)
5.119 D. Braichotte, H.v.d. Bergh: in ⌊5.26⌋, p. 183
5.120 W.E. Johnson, L.A. Schlie: Appl. Phys. Lett. 40, 798 (1982)
5.121 J.M. Gee, P.J. Hargis Jr., M.J. Carr, D.R. Tallant, R.W. Light: in ⌊5.16⌋, p. 15
5.122 A. Ishitani, M. Kanamori, H. Tsuya: J. Appl. Phys. 57, 2956 (1985)
5.123 D. Bäuerle: Laser und Optoelektronik 1, 29 (1985)
5.124 D. Bäuerle, G. Leyendecker, D. Wagner, E. Bauser, Y.C. Lu: Appl. Phys. A 30, 147 (1983)
5.125 T. Szörényi, G.Q. Zhang, Y.C. Du, R. Kullmer, D. Bäuerle: in ⌊5.53⌋
5.126 D. Bäuerle, P. Irsigler, G. Leyendecker, H. Noll, D. Wagner: Appl. Phys. Lett. 40, 819 (1982)
5.127 Y.C. Du, U. Kempfer, K. Piglmayer, D. Bäuerle, U.M. Titulaer: Appl. Phys. A 39, 167 (1986)
5.128 F. Petzoldt, S. Szikora, D. Bäuerle: unpublished
5.129 D.J. Ehrlich, R.M. Osgood, Jr., T.F. Deutsch: Appl. Phys. Lett. 39, 957 (1981)
5.130 I. Herman, F. Magnotta, D.E. Kotecki: J. Vac. Sci. Technol. A 4, 659 (1986)
5.131 F. Magnotta, I.P. Herman: Appl. Phys. Lett. 48, 195 (1986)
5.132 I.P. Herman, B.M. McWilliams, F. Mitlitsky, Hon Wah Chin, R.A. Hyde, L.L. Wood: in ⌊5.16⌋, p. 29
5.133 I.P. Herman: in ⌊5.26⌋, p. 396
5.134 B.M. McWilliams, I.P. Herman, F. Mitlitsky, R.A. Hyde, L.L. Wood: Appl. Phys. Lett. 43, 946 (1983)
5.135 R. Bilenchi, M. Musci: in *Chemical Vapor Deposition*, ed. by J.M. Blocher, G.E. Vuillard, G. Wahl, Electrochem. Soc. Proc. Ser. (Electrochemical Society, Pennington, NJ 1981) p. 275
5.136 C.P. Christensen, K.M. Lakin: Appl. Phys. Lett. 32, 254 (1978)
5.137 M. Hanabusa, A. Namiki, K. Yoshihara: Appl. Phys. Lett. 35, 626 (1979)
5.138 D. Tonneau, G. Auvert, Y. Pauleau: in ⌊5.18⌋, p. 125
5.139 Y. Pauleau, R. Stawski, Ph. Lami, G. Auvert: in ⌊5.16⌋, p. 41
5.140 D. Tonneau, G. Auvert, Y. Pauleau: in ⌊5.29⌋, p. 83
5.141 V. Baranauskas, C.I.Z. Mammana, R.E. Klinger, J.E. Greene: Appl. Phys. Lett. 36, 930 (1980)
5.142 M. Murahara, K. Toyoda: in ⌊5.26⌋, p. 252
5.143 T. Inoue, M. Konagai, K. Takahashi: Appl. Phys. Lett. 43, 774 (1983)
5.144 A. Yamada, M. Konagai, K. Takahashi: in ⌊5.29⌋, p. 29
5.145 Y. Mishima, M. Hirose, Y. Osaka, K. Nagamine, Y. Ashida, K. Isogaya: Jpn. J. Appl. Phys. 22, L46 (1983)
5.146 Y. Tarni, K. Aota, T. Sugiura, T. Saitoh: in ⌊5.16⌋, p. 109
5.147 T.D. Binnie, M.J. Colles, J.I.B. Wilson: in ⌊5.16⌋, p. 9
5.148 M. Hanabusa, S. Moriyama, H. Kikuchi: Thin Solid Films, 107, 227 (1983)
5.149 M. Hanabusa, H. Kikuchi, T. Iwanaga, K. Sugai: in ⌊5.26⌋, p. 197
5.150 M. Hanabusa, H. Kikuchi, T. Iwanaga, K. Sugai: in ⌊5.16⌋, p. 21
5.151 R. Bilenchi, I. Gianinoni, M. Musci: J. Appl. Phys. 53, 6479 (1982)
5.152 R. Bilenchi, I. Gianinoni, M. Musci: Appl. Phys. Lett. 47, 279 (1985)

226

5.153 R. Bilenchi, I. Gianinoni, M. Musci, R. Murri: in ⌊5.6⌋, p. 199
5.154 R. Bilenchi, M. Musci, R. Murri: in ⌊5.20⌋, p. 61
5.155 F. Curcio, I. Gianinoni, M. Musci: in ⌊5.53⌋
5.156 H.M. Branz, S. Fan, J.H. Flint, B.T. Fiske, D. Adler, J.S. Haggerty:
      Appl. Phys. Lett. 48, 171 (1986)
5.157 M. Meunier, T.R. Gattuso, D. Adler, J.S. Haggerty: Appl. Phys. Lett.
      43, 273 (1983)
5.158 T.R. Gattuso, M. Meunier, D. Adler, J.S. Haggerty: in ⌊5.6⌋, p. 215
5.159 S.A. Joyce, B. Roop, J.C. Schultz, K. Suzuki, J. Thoman,
      J.I. Steinfeld: in ⌊5.26⌋, p. 221
5.160 D. Tonneau, G. Auvert, Y. Pauleau: J. Chem. Phys., in press
5.161 Y. Pauleau, D.Tonneau, G. Auvert: in ⌊5.26⌋, p. 215
5.162 M. Hanabusa, M. Suzuki, S. Nishigaki: Appl. Phys. Lett. 38,
      385 (1981)
5.163 M. Hanabusa, M. Suzuki: Appl. Phys. Lett. 39, 431 (1981)
5.164 P.K. Boyer, K.A. Emery, H. Zarnani, G.J. Collins: Appl. Phys. Lett.
      45, 979 (1984)
5.165 A. Tabe, K. Jinguji, T. Yamada, N. Takato: Appl. Phys. A 38, 221 (1985)
5.166 P.K. Boyer, G.A. Roche, W.H. Ritchie, G.J. Collins: Appl. Phys. Lett.
      40, 716 (1982)
5.167 P.K. Boyer, W.H. Ritchie, G.J. Collins: J. Electrochem. Soc. 129, 2155
      (1982)
5.168 K.A. Emery, L.R. Thompson, D. Bishop, H. Zarnani, P.K. Boyer,
      C.A. Moore, J.J. Rocca, G.J. Collins: in ⌊5.16⌋, p. 81
5.169 S. Nishino, H. Honda, J. Matsunami: Jpn. J. Appl. Phys. 25, L87 (1986)
5.170 S. Szikora, W. Kräuter, D. Bäuerle: Mater. Lett. 2, 263 (1984)
5.171 R.R. Krchnavek, H.H. Gilgen, R.M. Osgood, Jr.: J. Vac. Sci. Technol.
      B 2, 641 (1984)
5.172 G.A. West, A. Gupta: in ⌊5.16⌋, p. 61
5.173 M.R. Aylett, J. Haigh: in ⌊5.26⌋, p. 263
5.174 O. Tabata, S. Kimura, S. Tabata, R. Makabe, S. Matsuura: in ⌈5.135⌋,
      p. 272
5.175 J.Y. Tsao, R.A. Becker, D.J. Ehrlich, F.J. Leonberger: Appl. Phys.
      Lett. 42, 559 (1983)
5.176 J. Mazumder, S.D. Allen: in ⌊5.110⌋, p. 73
5.177 G.A. West, A. Gupta, K.W. Beeson: Appl. Phys. Lett. 47, 476 (1985)
5.178 G.A. West, A. Gupta, K.W. Beeson: J. Vac. Sci. Technol. A 3, 2277
      (1985)
5.179 A.E. Adams, M.L. Lloyd, S.L. Morgan, N.G. Davis in ⌈5.26⌋, p. 269
5.180 T.F. Deutsch, D.D. Rathman: Appl. Phys. Lett. 45, 623 (1984)
5.181 S.D. Allen: Laser Focus, May, 14, (1983)
5.182 R.S. Berg, D.M. Mattox: in *Chemical Vapor Deposition,* ed. by F.A.
      Glaski (American Nuclear Society, Hinsdale, Ill. 1972) p. 196
5.183 Y.S. Liu, C.P. Yakymyshyn, H.R. Philipp, H.S. Cole, L.M. Levinson:
      J. Vac. Sci. Technol. B 3, 144 (1985)
5.184 R. Solanki, G.J. Collins: Appl. Phys. Lett. 42, 662 (1983)
5.185 D.J. Ehrlich, R.M. Osgood,Jr., T.F. Deutsch: J. Vac. Sci. Technol.
      20, 738 (1982)
5.186 H. Sankur, J.T. Cheung: J. Vac. Sci. Technol. A 1, 1806 (1983)
5.187 See, e.g., L.I. Maissel, R. Glang (eds.): *Handbook of Thin Film Tech-
      nology,* (McGraw-Hill, New York 1970)
5.188 U.M. Titulaer, D. Bäuerle: to be published
5.189 S.D. Allen, R.Y. Jan, S.M. Mazuk, S.D. Vernon: J. Appl. Phys. 58,
      327 (1985)
5.190 K. Piglmayer, D. Bäuerle: unpublished
5.191 R.K. Chan, R. McIntosh: Can. J. Chem. 40, 845 (1962)
5.192 H.E. Carlton, J.H. Oxley: AIChE J. 12, 86 (1967)
5.193 A.J. Goosen, J.A. Van den Berg: J.S. Afr. Chem. Inst. 25, 370
      (1972)

5.194 J.P. Day, R.G. Pearson, F. Basolo: J. Am. Chem. Soc. 90, 6933 (1968)
5.195 D. Bäuerle: in ⌊5.18⌋, p. 103
5.196 J. Doppelbauer, D. Bäuerle: to be published
5.197 C.H.J. v.d. Brekel: Ph. D. Thesis, Eindhoven (1978)
5.198 S. Szikora, J. Otto, D. Bäuerle: unpublished
5.199 D.K. Flynn, J.I. Steinfeld, D.S. Sethi: J. Appl. Phys. 59, 3914 (1986)
5.200 R.Y. Jan, S.D. Allen: in ⌊5.20⌋, p. 71
5.201 W. Kräuter, D. Bäuerle: unpublished
5.202 For a recent review see H.M.v. Driel, J.E. Sipe, J.F. Young: J. Lumin. 30, 446 (1985)
5.203 H.M.v. Driel, J.E. Sipe, J.F. Young: in *High Excitation and Short Pulse Phenomena,* ed. by M.H. Pilkuhn (North-Holland, Amsterdam 1984) p. 446
5.204 Z. Guosheng, P.M. Fauchet, A.E. Siegmann: Phys. Rev. B. 26, 5366 (1982)
5.205 JANAF Thermochemical Tables, National Bureau of Standards, USA (1970)
5.206 G. Auvert, D. Bensahel, A. Georges, V.T. Nguyen, P. Henoc, F. Morin, P. Croissard: Appl. Phys. Lett. 38, 613 (1981)
5.207 D.A. Kurtze, W.v. Saarloos, J.D. Weeks: Phys. Rev. B 30, 1398 (1984)
5.208 D.E. Carlson: Solar Energy Mater. 3, 503 (1980)
5.209 Y. Hamakawa: Solar Energy Mater. 8, 101 (1982)
5.210 P.G. LeComber, A.J. Snell, K.D. Mackenzie, W.E. Spear: J. Phys. (Paris) C 4, 423 (1981)
5.211 B.K. Chakravertry, D. Kaplan (eds.): *Amorphous and Liquid Semicon-ductors,* J. Phys. (Paris) 42, C-4 (1981)
5.212 F. Evangelisti, J. Stuke (eds.): *Amorphous and Liquid Semiconductors I, II* (North-Holland, Amsterdam 1985)
5.213 M. Meunier, J.H. Flint, D. Adler, J.S. Haggerty: in ⌊5.16⌋, p. 397
5.214 B.A. Scott, J.A. Reiner, R.M. Plecenik, E.E. Simonyi, W. Reuter: Appl. Phys. Lett. 40, 973 (1982)
5.215 C.G. Newman, H.E. O'Neal, M.A. Ring, F. Leska, N. Skipley: Int. J. Chem. Kinet. 11, 1167 (1979)
5.216 K.J. Sladek: J. Eletrochem. Soc. 118, 645 (1971)
5.217 N. Pütz, H. Heinecke, E. Veuhoff, G. Arens, M. Heyen, H. Lüth, P. Balk: J. Cryst. Growth 68, 194 (1984)
5.218 See e.g., H. Sankur: in ⌊5.20⌋, p. 78
5.219 M.Hanabusa, S. Moriyama, H. Kikuchi: Thin Solid Films 107, 227 (1983)
5.220 R.S. Novicki: J. Vac. Sci. Technol. 14, 127 (1977)
5.221 V.P. Ageev, S. Allen, V.I. Konov, V.I. Melnikov, H.P. Preiswerk: in ⌊5.53⌋
5.222 M.H. Gelchinski, R.J. von Gutfeld, L.T. Romankiw: in *Interfaces under Laser Irradiation,* ed. by D. Bäuerle, L.D. Laude, M. Wautelet, Nato ASI Series (M. Nijhoff, Dordrecht 1987)
5.223 L.D. Laude: in ⌊5.26⌋, p. 355
5.224 T. Szörényi, G.Q. Zhang, D. Bäuerle: to be published
5.225 N. Itoh: in ⌊5.222⌋
5.226 S.M. Metev: in ⌊5.53⌋
5.227 K. Kumata, U. Itoh, Y. Toyoshima, N. Tanaka, H. Anzai, A. Matsuda: Appl. Phys. Lett. 48, 1380 (1986)
5.228 H. Yokoyama, S. Kishida, K. Washio: Appl. Phys. Lett. 44, 755 (1984)
5.229 B.J. Morris: Appl. Phys. Lett. 48, 867 (1986)
5.230 T.R. Jervis: J. Appl. Phys. 58, 1400 (1985)

Chapter 6

6.1 I. Ursu, I. Apostol, M. Dinescu, A. Hening, I.N. Mihailescu, A.M. Prokhorov, N.I. Chapliev, V.I. Konov: J. Appl. Phys. 58, 1765 (1985)

6.2 I. Ursu, I. Apostol, I.N. Mihailescu, L.C. Nistor, V.S. Teodorescu,
     A.M. Prokhorov, V.I. Konov, N.I. Chapliev: in *Surface Studies with
     Lasers,* ed. by F.R. Aussenegg, A. Leitner, M.E. Lippitsch, Springer
     Ser. Chem. Phys., Vol. 33 (Springer, Berlin, Heidelberg 1983) p. 234
6.3 R. Andrew, L. Baufay, A. Pigeolet, M. Wautelet: in *Laser Diagnostics
     and Photochemical Processing for Semiconductor Devices,* ed. by R.M.
     Osgood, S.R.J. Brueck, H.R. Schlossberg (North-Holland, New York 1983)
     p. 283
6.4 R. Andrew, L. Baufay, A. Pigeolet, M. Wautelet: in ⌊6.117⌋, p. 163
6.5 M. Wautelet, L. Baufay: Thin Solid Films 100, L9 (1983)
6.6 M. Wautelet: Mater. Lett. 2, 20 (1983)
6.7 S.M. Metev, S.K. Savtchenko, K.V. Stamenov, V.P. Veiko, G.A. Kotov,
     G.D. Shandibina: IEEE J. QE-17, 2004 (1981)
6.8 S.M. Metev, S.K. Savtchenko, K.V. Stamenov, V.P. Veiko, G.A. Kotov,
     G.D. Shandibina: in *Laser Assisted Modification and Synthesis of
     Materials,* Proc. Int. Winter School, Gyuletchitza, Bulgaria 1985, ed.
     by S. Metev, p. 61
6.9 S.M. Metev, S.K. Savtchenko, K.V. Stamenov, V.P. Veiko, G.A. Kotov,
     G.D. Shandibina: J. Phys. D 13, L 75 (1980)
6.10 M.J. Birjega, C.A. Constantin, M. Dinescu, I.T. Florescu,
     I.N. Mihailescu, L. Nanu, N. Popescu, C. Sarbu: in *Laser Controlled
     Chemical Processing of Surfaces,* ed. by A.W. Johnson, D.J. Ehrlich,
     H.R. Schlossberg (North-Holland, New York 1984) p. 289
6.11 I. Ursu, L.C. Nistor, V.S. Teodorescu, I.N. Mihailescu, I. Apostol,
     L. Nanu, A.M. Prokhorov, N.I. Chapliev, V.I. Konov, V.N. Tokarev,
     V.G. Ralchenko: SPIE 398, 398 (1983)
6.12 I. Ursu, L.C. Nistor, V.S. Teodorescu, I.N. Mihailescu, I. Apostol,
     L. Nanu, A.M. Prokhorov, N.I. Chapliev, V.I. Konov, V.N. Tokarev,
     V.G. Ralchenko: J. Phys. (Paris), Lett. 45, L737 (1984)
6.13 I. Ursu, L.C. Nistor, V.S. Teodorescu, I.N. Mihailescu, I. Apostol,
     L. Nanu, A.M. Prokhorov, N.I. Chapliev, V.I. Konov, V.N. Tokarev,
     V.G. Ralchenko: Appl. Phys. Lett. 44, 188 (1984)
6.14 I. Ursu, L.C. Nistor, V.S. Teodorescu, I.N. Mihailescu, I. Apostol,
     L. Nanu, A.M. Prokhorov, N.I. Chapliev, V.I. Konov, V.N. Tokarev,
     V.G. Ralchenko: Appl. Phys. A 29, 209 (1982)
6.15 I. Ursu, L.C. Nistor, V.S. Teodorescu, I.N. Mihailescu, I. Apostol,
     L. Nanu, A.M. Prokhorov, N.I. Chapliev, V.I. Konov, V.N. Tokarev,
     V.G. Ralchenko: in *Laser Processing and Diagnostics II,* ed. by
     D. Bäuerle, K.L. Kompa, L.D. Laude (Physique, Les Ulis 1986)
6.16 D.Braichotte, K. Ernst, R. Monot, J.M. Philippoz, M. Qiu, H.v.d. Bergh:
     Frühjahrstagung der Schweiz 58, 879 (1985)
6.17 J. Siejka, R. Srinivasan, J. Perrière, S. Lazare: Appl. Phys. Lett.
     (1986)
6.18 J. Siejka, R. Srinivasan, J. Perrière: in ⌊6.117⌋, p. 139
6.19 J. Siejka, J. Perrière, R. Srinivasan: Appl. Phys. Lett. 46, 773 (1985)
6.20 M. Fukuda, K. Takahei: J. Appl. Phys. 57, 129 (1985)
6.21 W.G. Petro, I. Hino, S. Eglash, I. Lindau, C.Y. Su, W.E. Spicer: J.
     Vac. Sci. Technol. 21, 405 (1982)
6.22 K.A. Bertness, W.G. Petro, J.A. Silberman, D.J. Friedman,
     T. Kendelewicz, W.E. Spicer: in *Laser Chemical Processing of Semicon-
     ductor Devices,* Extended Abstracts, ed. by F. A. Houle, T.F. Deutsch,
     R.M. Osgood (Materials Research Society, Boston 1984) p. 129
6.23 C. Cohen, J. Siejka, M. Berti, A.V. Drigo, G.G. Bentini, D. Pribat,
     E. Jannitti: J. Appl. Phys. 55, 4081 (1984)
6.24 G.G. Bentini, M. Berti, C. Cohen, A.V. Drigo, E. Jannitti, D. Pribat,
     J. Siejka: J. Phys. (Paris) 43, C1-229 (1982)
6.25 M. Matsuura, M. Ishida, A. Suzuki, K. Hara: Jpn. J. Appl. Phys. 20,
     L 726 (1981)

6.26 H. Zarnani, P.K. Boyer, M. Fathipour, C.W. Wilmsen, R. Solanki, G.J. Collins: in ⌊6.22⌋, p. 122
6.27 H. Zarnani, P.K. Boyer, M. Fathipour, C.W. Wilmsen, R. Solanki, G.J. Collins: J. Appl. Phys. (1986)
6.28 R.F. Marks, R.A. Pollak, Ph. Avouris, C.T. Lin, Y.J. Théfaine: J. Chem. Phys. $\underline{78}$, 4270 (1983)
6.29 R.F. Marks, R.A. Pollak, Ph. Avouris, C.T. Lin, Y.J. Théfaine: in ⌊6.3⌋, p. 257
6.30 K. Nakamura, M. Hikita, H. Asano, A. Terada: Jpn. J. Appl. Phys. $\underline{21}$, 672 (1982)
6.31 S.E. Blum, K. Brown, R. Srinivasan: Appl. Phys. Lett. $\underline{43}$, 1026 (1983)
6.32 E. Fogarassy: in ⌊6.117⌋, p. 153
6.33 Y.S. Liu, S.W.Chiang, F. Bacon: Appl. Phys. Lett. $\underline{38}$, 1005 (1981)
6.34 T.E. Orlowski, H. Richter: Appl. Phys. Lett. $\underline{45}$, $\overline{241}$ (1984)
6.35 H. Richter, T.E. Orlowski, M. Kelly, G. Margaritondo: J. Appl. Phys. $\underline{56}$, 2351 (1984)
6.36 E.M. Young, W.A. Tiller: Appl. Phys. Lett. $\underline{42}$, 63 (1983)
6.37 S.A. Schafer, S.A. Lyon: J. Vac. Sci. Technol. $\underline{19}$, 494 (1981)
6.38 S.A. Schafer, S.A. Lyon: J. Vac. Sci. Technol. $\underline{21}$, 422 (1982)
6.39 I.W. Boyd: Appl. Phys. Lett. $\underline{42}$, 728 (1983)
6.40 J.F. Gibbons: Jpn. J. Appl. Phys. Suppl. $\underline{19}$, 121 (1981)
6.41 A. Cros, F. Salvan, J. Derrien: Appl. Phys. A $\underline{28}$, 241 (1982)
6.42 G.G. Bentini, M. Berti, A.V. Drigo, E. Jannitti, C. Cohen, J. Siejka: in ⌊6.22⌋, p. 126
6.43 M. Berti, A.V. Drigo, G.G. Bentini, C. Cohen, J. Siejka, E. Jannitti: in ⌊6.117⌋, p. 131; C. Cohen, J. Siejka, G.G. Bentini, M. Berti, L.F. Dona Dalle Rose, A.V. Drigo: in ⌊6.15⌋
6.44 A. Garulli, M. Servidori, I. Vecchi: J. Phys. D. $\underline{13}$, L199 (1981)
6.45 S.W. Chiang, Y.S. Liu, R.F. Reihl: Appl. Phys. Lett. $\underline{39}$, 752 (1981)
6.46 K. Hoh, H. Koyama, K. Uda, Y. Miura: Jpn. J. Appl. Phys. $\underline{19}$, L375 (1980)
6.47 T.D. Binnie, M.J. Colles, J.I.B. Wilson: in ⌊6.10⌋, p. 9
6.48 I.W. Boyd, J.I.B. Wilson: Appl. Phys. Lett. $\underline{41}$, 162 (1982)
6.49 I.W. Boyd, J.I.B. Wilson: Solid-State Electron. $\underline{27}$, 209 (1984)
6.50 I.W. Boyd: Appl. Phys. A $\underline{31}$, 71 (1983)
6.51 I.W. Boyd, J.I.B. Wilson, J.L. West: Thin Solid Films, 83 L173 (1981)
6.52 T. Sugii, T. Ito, H. Ishikawa: Appl. Phys. Lett. $\underline{45}$, 966 (1984)
6.53 M. Wautelet, L. Baufay, M.C. Joliet, R. Andrew: Thin Solid Films $\underline{129}$, L67 (1985)
6.54 I. Ursu, I.N. Mihailescu, A.M. Prokhorov, V.I. Konov: Appl. Phys. A (1986)
6.55 I. Ursu, I.N. Mihailescu, L. Nanu, A.M. Prokhorov, V.I. Konov, V.G. Ralchenko: Appl. Phys. Lett. $\underline{46}$, 110 (1985)
6.56 I. Ursu, I.N. Mihailescu, L.C. Nistor, A. Popa, M. Popescu, V.S. Teodorescu, L. Nanu, A.M. Prokhorov, V.I. Konov, V.N. Tokarev: J. Phys. D $\underline{18}$, 1693 (1985)
6.57 R. Merlin, T.A. Perry: Appl. Phys. Lett. $\underline{45}$, 852 (1984)
6.58 T. Szörényi, L. Baufay, M.C. Joliet, F. Hanns, R. Andrew, I. Hevesi: Appl. Phys. A $\underline{39}$, 251 (1986)
6.59 I. Ursu, L. Nanu, M. Dinescu, A. Hening, I.N. Mihailescu, L.C. Nistor, V.S. Teodorescu, E. Szil, I. Hevesi, J. Kovacs, L. Nanai: Appl. Phys. A $\underline{35}$, 103 (1984)
6.60 I. Ursu, I.N. Mihailescu, L.C. Nistor, M. Popescu, V.S. Teodorescu, A.M. Prokhorov, V.G. Ralchenko, V.I. Konov: J. Appl. Phys. (1986)
6.61 N. Cabrera, N.F. Mott: Rep. Prog. Phys. $\underline{12}$, 163 (1949)
6.62 F.P. Fehner, N.F. Mott: J. Oxidation Met. $\overline{2}$, 59 (1970)
6.63 I.W. Boyd: in *Laser Processing and Diagnostics*, ed. by D. Bäuerle, Springer Ser. Chem. Phys., Vol. 39 (Springer, Berlin, Heidelberg 1984) p. 274; in ⌊6.15⌋

6.64   T. Tokuyama, S. Kimura, T. Warabisako, E. Murakami, K. Miyake:
        in ⌊6.63⌋, p. 288
6.65   H.F. Winters, J.W. Coburn, T.J. Chuang: J. Vac. Sci. Technol. A 1,
        1157 (1983)
6.66   H.F. Winters, J.W. Coburn, T.J. Chuang: J. Vac. Sci. Technol. B 1, 469
        (1983)
6.67   C. Fiori: Phys. Rev. Lett. 52, 2077 (1984)
6.68   C. Fiori, R.A.B. Devine: Phys. Rev. Lett. 52, 2081 (1984)
6.69   D. Bäuerle: Laser und Optoelektronik 1, 29 (1985)
6.70   J. Otto, R. Stumpe, D. Bäuerle: in ⌊6.63⌋, p. 320
6.71   M. Eyett, D. Bäuerle: unpublished
6.72   A. Kapenieks, M. Eyett, D. Bäuerle: Appl. Phys. A 41 (1986)
6.73   A. Kapenieks, R. Stumpe, M. Eyett, D. Bäuerle: in IEEE Int. Symp. on
        Applications of Ferroelectrics (1986)
6.74   See e.g., Landolt-Börnstein, New Series. Group 3, Vol. 16, Part a.
        Ferroelectrics and Related Substances: Oxides (Springer, Berlin,
        Heidelberg 1981)
6.75   G. Perluzzo, J. Destry: Can. J. Phys. 56, 453 (1978)
6.76   D. Bäuerle, D. Wagner, M. Wöhlecke, B. Dorner, H. Kraxenberger: Z.
        Phys. B 38, 335 (1980)
6.77   D. Bäuerle: Ferroelectrics 20, 555 (1978)
6.78   D. Wagner, D. Bäuerle, F. Schwabl, B. Dorner, H. Kraxenberger: Z. Phys.
        B 37, 317 (1980)
6.79   D. Bäuerle, W. Rehwald: Solid State Commun. 27, 1343 (1978)
6.80   R. Migoni, H. Bilz, D. Bäuerle: Phys. Rev. Lett. 37, 1155 (1976)
6.81   R. Migoni, H. Bilz, D. Bäuerle: in *Lattice Dynamics*, ed. by M. Balkanski
        (Flamarion Press 1978) p. 650
6.82   R. Migoni, H. Bilz, D. Bäuerle: Ferroelectrics 20, 157 (1978)
6.83   A. Bussmann-Holder, H. Bilz, D. Bäuerle, D. Wagner: Z. Phys. B 41,
        353 (1981)
6.84   D. Bäuerle: in ⌊6.63⌋, p. 166
6.85   T.F. Deutsch, J.C.C. Fan, D.J. Ehrlich, G.W. Turner, R.L. Chapman,
        R.P. Gale: Appl. Phys. Lett. 40, 722 (1982)
6.86   T.F. Deutsch, D.J. Ehrlich, D.D. Rathman, D.J. Silversmith, R.M. Osgood:
        Appl. Phys. Lett. 39, 825 (1981)
6.87   T.F. Deutsch, D.J. Ehrlich, D.D. Rathman, D.J. Silversmith, R.M. Osgood:
        in ⌊6.3⌋, p. 225
6.88   H. Beneking: in ⌊6.63⌋, p. 188
6.89   H. Kräutle, W. Roth, A. Krings, H. Beneking: in ⌊6.10⌋, p. 353
6.90   H. Kräutle, P. Roentgen, M. Maier, H. Beneking: Appl. Phys. A 38, 49
        (1985); H. Kräutle, D. Wachenschwanz: Solid-State Electron. 28, 601
        (1985)
6.91   T.F. Deutsch, D.J. Ehrlich, R.M. Osgood, Z.L. Liau: Appl. Phys. Lett.
        36, 847 (1980)
6.92   D.J. Ehrlich, R.M. Osgood, T.F. Deutsch: Appl. Phys. Lett. 36,
        916 (1980)
6.93   T.F. Deutsch, J.C.C. Fan, G.W. Turner, R.L. Chapman, D.J. Ehrlich,
        R.M. Osgood: Appl. Phys. Lett. 38, 144 (1981)
6.94   E. Fogarassy, R. Stuck, J.J. Grob, P. Siffert: J. Appl. Phys. 52,
        1076 (1981)
6.95   F.E. Harper, M.I. Cohen: Solid-State Electron. 13, 1103 (1970)
6.96   D.J. Ehrlich, J.Y. Tsao: Appl. Phys. Lett. 41, 297 (1982)
6.97   T.F. Deutsch: in ⌊6.63⌋, p. 239
6.98   K.G. Ibbs, M.L. Lloyd: Opt. Laser Technol.,February (1984)
6.99   K.G. Ibbs, M.L. Lloyd: in ⌊6.3⌋, p. 243
6.100   J. Narayan, R.T. Young, R.F. Wood, W.H. Christie: Appl. Phys. Lett.
        33, 338 (1978)
6.101   R.T. Young, R.F. Wood, J. Narayan, C.W. White, W.H. Christie: IEEE
        Trans. ED-27, 807 (1980)

6.102 G.B. Turner, D. Tarrant, G. Pollock, P. Pressley, R. Press: Appl. Phys. Lett. 39, 967 (1981)

6.103 R. Stuck, E. Fogarassy, J.C. Muller, M. Hodeau, A. Wattieux, P. Siffert: Appl. Phys. Lett. 38, 715 (1981)

6.104 J.C. Muller, P. Siffert, C.T. Ho, J.I. Hanoka, F.V. Wald: J. Phys. (Paris) 43, C1-235 (1982)

6.105 E.P. Fogarassy, D.H. Lowndes, J. Narayan, C.W. White: J. Appl. Phys. 58, 2167 (1985)

6.106 E. Fogarassy, R. Stuck, M. Toulemonde, D. Salles, P. Siffert: J. Appl. Phys. 54, 5059 (1983)

6.107 J.A. McKay, J.T. Schriempf: IEEE J. QE-10, 2008 (1981)

6.108 J. Götzlich, H. Ryssel: in ⌊6.63⌋, p. 40

6.109 S.D. Ferries, H.J. Leamy, J.M. Poate (eds.): *Laser-Solid Interactions and Laser Processing* (AIP, New York 1979)

6.110 C.W. White, P.S. Peercy (eds.): *Laser and Electron Beam Processing of Materials* (Academic, New York 1980)

6.111 J.F. Gibbons, L.D. Hess, T.W. Sigmon (eds.): *Laser and Electron Beam Solid Interactions and Materials Processing* (North-Holland, New York 1981)

6.112 J.M. Poate, J.W. Mayer (eds.): *Laser Annealing of Semiconductors* (Academic, New York 1982)

6.113 J. Narayan, W.L. Brown, R.A. Lemons (eds.): *Laser-Solid Interactions and Transient Thermal Processing of Materials* (North-Holland, New York 1983)

6.114 A.G. Cullis: Rep. Prog. Phys. 48, 1155 (1985)

6.115 B.R. Appleton, G.K. Celler: *Laser and Electron-Beam Interactions with Solids* (North-Holland, New York 1982)

6.116 K. Hennig (ed.): *Energy Pulse Modification of Semiconductors and Related Materials* (Akademie der Wissenschaften, Dresden 1984)

6.117 V.T. Nguyen, A.G. Cullis: *Energy Beam-Solid Interactions and Transient Thermal Processing* (Physique, Les Ulis 1985)

6.118 I. Ursu, A.M. Prokhorov (eds.): *Trends in Quantum Electronics,* Abstracts (European Physical Society, Bucharest 1985)

6.119 I.W. Boyd: in *Interfaces under Laser Irradiation,* ed. by D. Bäuerle, L.D. Laude, M. Wautelet, Nato ASI Series (M. Nijhoff, Dordrecht 1987)

6.120 A. Kapenieks, R. Stumpe, M. Eyett, D. Bäuerle: in ⌊6.15⌋

6.121 Emulsitone and XB-100 are supplied by Allied Chemical, USA

6.122 K. Horioka, H. Okano, M. Sekine, Y. Horiike: in Proc. *Dry Process Symposium* (Inst. Electr. Eng., Tokyo, 1984) p. 80

6.123 M. Thuillard, M.v. Allmen: in ⌊6.15⌋; Appl. Phys. Lett. 47, 936 (1985); ibid. 48, 1045 (1986)

6.124 A.G. Akimov, A.P. Gagarin, V.G. Dagurov, V.S. Makin, S.D. Pudkov: Sov. Phys. Tech. Phys. 25, 1439 (1980)

6.125 P.G. Carey, T.W. Sigmon, R.L. Press, T.S. Fahlen: IEEE Electron Device Lett. 6, 291 (1985)

Chapter 7

7.1 M.v. Allmen (ed.): *Amorphous Metals and Nonequilibrium Processing* (Physique, Les Ulis 1984)

7.2 M.v. Allmen: in *Energy Beam-Solid Interactions and Transient Thermal Processing,* ed. by V.T. Nguyen, A.G. Cullis (Physique, Les Ulis 1985) p. 373

7.3 M.v. Allmen: in *Laser Processing and Diagnostics II,* ed. by D. Bäuerle, K.L. Kompa, L.D. Laude (Physique, Les Ulis 1986)

7.4 Z.L. Liau, B.Y. Tsaur, J.W. Mayer: Appl. Phys. Lett. 34, 221 (1979)

7.5 T. Shibata, J.F. Gibbons, T.W. Sigmon: Appl. Phys. Lett. 36, 566 (1980)

7.6 J.M. Poate, H.J. Leamy, T.T. Sheng, G.K. Celler: Appl. Phys. Lett. 33,
     918 (1978)
7.7 B.M. Ditchek, T. Emma: Appl. Phys. Lett. 45, 955 (1984)
7.8 E. D'Anna, G. Leggieri, A. Luches, M.R. Perrone, G. Majni, I. Catalano:
     in Laser Processing and Diagnostics, ed. by D. Bäuerle, Springer Ser.
     Chem. Phys., Vol. 39 (Springer, Berlin, Heidelberg 1984) p. 370
7.9 J. Narayan, D. Fathy, O.W. Holland, B.R. Appleton, R.F. Davis,
     P.F. Becher: J. Appl. Phys. 56, 1577 (1984)
7.10 J. Narayan, D. Fathy, O.W. Holland, B.R. Appleton, R.F. Davis,
     P.F. Becher: Mater. Lett. 3, 261 (1985)
7.11 M.G. Grimaldi, P. Baeri, E. Rimini, G. Celotti: Appl. Phys. Lett. 43,
     244 (1983)
7.12 P. Baeri, S.U. Campisano, F. Priolo, E. Rimini: in ⌊7.2⌋, p. 237
7.13 M.v. Allmen, S.S. Lau, M. Mäenpää, B.Y. Tsaur: Appl. Phys. Lett.
     36, 205 (1980)
7.14 M.v. Allmen, M. Wittmer: Appl. Phys. Lett. 34, 68 (1979)
7.15 G.G. Bentini, M. Servidori, C. Cohen, R. Nipoti, A.V. Drigo:
     J. Appl. Phys. 53, 1525 (1982)
7.16 G.J.v. Gurp, G.E.J. Eggermont, Y. Tamminga, W.T. Stacy,
     J.R.M. Gijsbers: Appl. Phys. Lett. 35, 273 (1979)
7.17 E. D'Anna, G. Leggieri, A. Luches, G. Majni, G. Ottaviani,
     M.R. Perrone: in ⌊7.2⌋, p. 187
7.18 A.E. Adams, M.L. Lloyd, S.L. Morgan, N.G. Davis in ⌊7.8⌋, p. 269
7.19 K.C.R Chiu, J.M. Poate, L.C. Feldman, C.J. Doherty: Appl. Phys. Lett.
     36, 544 (1980)
7.20 K.C.R. Chiu, J.M. Poate, J.E. Rowe, T.T. Sheng, A.G. Cullis: Appl. Phys.
     Lett. 38, 988 (1982)
7.21 L.D. Laude: in ⌊7.8⌋, p. 355
7.22 L.D. Laude, M. Wautelet, R. Andrew: Appl. Phys. A 40, 133 (1986)
7.23 R. Andrew, M. Ledezma, M. Lovato, M. Wautelet, L.D. Laude:
     Appl. Phys. Lett. 35, 418 (1979)
7.24 R. Andrew, L. Baufay, A. Pigeolet, L.D. Laude: J. Appl. Phys. 53,
     4862 (1982)
7.25 L. Baufay, A. Pigeolet, L.D. Laude: J. Appl. Phys. 54, 660 (1983)
7.26 L. Baufay, R. Andrew, A. Pigeolet, L.D. Laude: Rev. Phys. Appl. 18,
     207 (1983)
7.27 L. Baufay, R. Andrew, A. Pigeolet, L.D. Laude: Physica 117 B, 1027
     (1983)
7.28 R. Andrew, L. Baufay, Y. Canivez, A. Pigeolet: J. Phys. (Paris) 44,
     C5-495 (1983)
7.29 L. Baufay, D. Dispa, A. Pigeolet, L.D. Laude: J. Cryst. Growth 59, 143
     (1982)
7.30 L.D. Laude: in Interfaces under Laser Irradiation, ed. by D. Bäuerle,
     L.D. Laude, M. Wautelet, Nato ASI Series (M. Nijhoff, Dordrecht 1987)
7.31 L.D. Laude, M. Wautelet: Mater. Lett. 2, 280 (1984)
7.32 C. Antoniadis, M.C. Joliet: Thin Solid Films 115, 75 (1984)
7.33 C. Antoniadis, L.D. Laude, P. Pierrard: in ⌊7.2⌋, p. 387
7.34 M.C. Joliet, C. Antoniadis, R. Andrew, L.D. Laude: in ⌊7.8⌋, p. 366
7.35 M.C. Joliet, C. Antoniadis, R. Andrew, L.D. Laude: Appl. Phys. Lett.
     46, 266 (1985)
7.36 M.C. Joliet, C. Antoniadis, L.D. Laude: Thin Solid Films 126, 143 (1985)
7.37 S. Oguz, W. Paul, T.F. Deutsch, B.Y. Tsaur, D.V. Murphy: Appl. Phys.
     Lett. 43, 848 (1983)
7.38 M. Wautelet: Phys. Lett. 108 A, 99 (1985)
7.39 E. D'Anna, G. Leggieri, A. Luches, G. Majni, F. Nava, G. Ottaviani:
     Appl. Phys. A 40, 183 (1986)

Chapter 8

8.1 T.J. Chuang, I. Hussla, W. Sesselmann: in *Laser Processing and Diagnostics,* ed. by D. Bäuerle, Springer Ser. Chem. Phys., Vol. 39 (Springer, Berlin, Heidelberg 1984) p. 300

8.2 W. Sesselmann, T.J. Chuang: J. Vac. Sci. Technol. B 3, 1507 (1985)

8.3 W. Sesselmann, T.J. Chuang: Surf. Sci. 162, 1007 (1985)

8.4 G. Koren, F.Ho, J.J. Ritsko: Appl. Phys. Lett. 46, 1006 (1985)

8.5 G. Koren, F.Ho, J.J. Ritsko: Appl. Phys. A 40, 13 (1986)

8.6 J.Y. Tsao, D.J. Ehrlich: Appl. Phys. Lett. 43, 146 (1983)

8.7 D.J. Ehrlich, R.M. Osgood, T.F. Deutsch: Appl. Phys. Lett. 38, 399 (1981)

8.8 T.J. Chuang: in *Laser Controlled Chemical Processing of Surfaces,* ed. by A.W. Johnson, D.J. Ehrlich, H.R. Schlossberg (North-Holland, New York 1984) p. 185

8.9 T.J. Chuang: in *Laser Diagnostics and Photochemical Processing for Semiconductor Devices,* ed. by R.M.Osgood, S.R.J. Brueck, H.R. Schlossberg (North-Holland, New York 1983) p. 45

8.10 R.J. von Gutfeld: in ⌊8.1⌋, p. 323

8.11 R.J. von Gutfeld: Denki Kagaku 52, 452 (1984)

8.12 R.J. von Gutfeld, R.T. Hodgson: Appl. Phys. Lett. 40, 352 (1982)

8.13 A. Kapenieks, R. Stumpe, M. Eyett, D. Bäuerle: in *Laser Processing and Diagnostics II,* ed. by D. Bäuerle, K.L. Kompa, L.D. Laude (Physique, Les Ulis 1986)

8.14 M. Eyett, D. Bäuerle, W. Wersing, K. Lubitz, H. Thomann: Appl. Phys. A 40, 235 (1986)

8.15 G. Koren, J.T.C. Yeh: J. Appl. Phys. 56, 2120 (1984)

8.16 C.I.H. Ashby: in ⌊8.8⌋, p. 173

8.17 R.M. Osgood, A. Sanchez-Rubio, D.J. Ehrlich, V. Daneu: Appl. Phys. Lett. 40, 391 (1983)

8.18 R. Tenne, V. Marcu, Y. Prior: Appl. Phys. A 37, 205 (1985)

8.19 V.A. Tyagai, V.A. Sterligov, G.Y. Kolbasov: Electrochim. Acta, 22, 819 (1977)

8.20 C. Uzan, R. Legros, Y. Marfaing, R. Triboulet: Appl. Phys. Lett. 45, 879 (1984)

8.21 T. Donohue: in ⌊8.1⌋, p. 332

8.22 T. Donohue: Proc. SPIE 482, 125 (1984)

8.23 J.C. Puippe, R.E. Acosta, R.J. von Gutfeld: J. Electrochem. Soc. 128, 2539 (1981)

8.24 J.C. Puippe, R.E. Acosta, R.J. von Gutfeld: Oberfläche-Surf. 22, 294 (1981)

8.25 K. Ando, N. Takeda, N. Koshizuka: Appl. Phys. Lett. 46, 1107 (1985)

8.26 P. Brewer, S. Halle, R.M. Osgood: Appl. Phys. Lett. 45, 475 (1984)

8.27 P. Brewer, S. Halle, R.M. Osgood: in ⌊8.8⌋, p. 179

8.28 P. Brewer, D. McClure, R.M. Osgood: in *Laser Chemical Processing of Semiconductor Devices,* Extended Abstracts, ed. by F.A. Houle, T.F. Deutsch, R.M. Osgood (Materials Research Society, Boston 1984) p. 102

8.29 M. Hirose, S. Yokoyama, Y. Yamakage: J. Vac. Sci. Technol. B 3, 1445 (1985)

8.30 M. Hirose, S. Yokoyama, Y. Yamakage: Appl. Phys. Lett. 47, 389 (1985)

8.31 D.J. Ehrlich, R.M. Osgood, T.F. Deutsch: Appl. Phys. Lett. 36, 698 (1980)

8.32 A.W. Tucker, M. Birnbaum: IEEE Electron Device Lett. 4, 39 (1983)

8.33 M. Takai, H. Nakai, J. Tsuchimoto, J. Tokuda, T. Minamisono, K. Gamo, S. Namba: in ⌊8.1⌋, p. 315

8.34 M. Takai, H. Nakai, J. Tsuchimoto, J. Tokuda, T. Minamisono, K. Gamo, S. Namba: in ⌊8.8⌋, p. 211

8.35 M. Takai, H. Nakai, J. Tsuchimoto, J. Tokuda, T. Minamisono, K. Gamo, S. Namba: Jpn. J. Appl. Phys. 22, L 757 (1983); ibid. 23, L 852 (1984); ibid. 24, L 705, 755 (1985)

8.36 M. Takai, J. Tsuchimoto, H. Nakai, J. Tokuda, K. Gamo, S. Namba: in *Beam Induced Chemical Processes,* Extended Abstracts, ed. by R.J. von Gutfeld, J.E. Greene, H. Schlossberg (Materials Research Society, Boston 1985) p. 129

8.37 N. Tsukada, S. Semura, H. Saito, S. Sugata, K. Asakawa, Y. Mita: J. Appl. Phys. 55, 3417 (1984)

8.38 D.V. Podlesnik, H.H. Gilgen, R.M. Osgood: Appl. Phys. Lett. 45, 563 (1984)

8.39 F. Kuhn-Kuhnenfeld: J. Electrochem. Soc. 119, 1063 (1972)

8.40 D.V. Podlesnik, H.H. Gilgen, R.M. Osgood, A. Sanchez, V. Daneu: Appl. Phys. Lett. 43, 1083 (1983)

8.41 D.V. Podlesnik, H.H. Gilgen, R.M. Osgood, A. Sanchez, V. Daneu: in ⌊8.9⌋, p. 57

8.42 Y. Aoyagi, S. Masuda, A. Doi, S. Namba: Jpn. J. Appl. Phys. 24, L 294 (1985)

8.43 D. Sun, F. Li, Y.C. Du: in Proc. Int. Conf. Lasers, China 1983, p. 582

8.44 G.C. Tisone, A.W. Johnson: Appl. Phys. Lett. 42, 530 (1983)

8.45 G.C. Tisone, A.W. Johnson: in ⌊8.9⌋, p. 73; in ⌊8.1⌋, p. 337

8.46 N. Tsukada, S. Sugata, H. Saitoh, Y. Mita: Appl. Phys. Lett. 43, 189 (1983)

8.47 R.W. Haynes, G.M. Metze, V.G. Kreismanis, L.F. Eastman: Appl. Phys. Lett. 37, 344 (1980)

8.48 S. Mottet, B. Henry: Electron. Lett. 19, 919 (1983)

8.49 F.W. Ostermayer, P.A. Kohl: Appl. Phys. Lett. 39, 76 (1981)

8.50 R.P. Salathé, C. Rao: in ⌊8.9⌋, p. 65

8.51 C.I.H. Ashby: Appl. Phys. Lett. 45, 892 (1984)

8.52 C.I.H. Ashby, R.M. Biefeld: Appl. Phys. Lett. 47, 62 (1985)

8.53 B. Zysset, R.P. Salathé, J.L. Martin, R. Gotthardt, F.K. Reinhart: in ⌊8.1⌋, p. 469

8.54 B. Zysset, R.P. Salathé, J.L. Martin, R. Gotthardt, F.K. Reinhart: Appl. Phys. Lett. 45, 428 (1984)

8.55 I.M. Beterov, V.P. Chebotaev, N.I. Yurshina, B. Ya Yurshin: Sov. J. Quantum Electron. 8, 1310 (1978)

8.56 G.P. Davis, C.A. Moore, R.A. Gottscho: in *Laser Assisted Deposition, Etching and Doping,* ed. by S.D. Allen, Proc. SPIE Conf. 459, 115 (1984)

8.57 G.P. Davis, C.A. Moore, R.A. Gottscho: J. Appl. Phys. 56, 1808 (1984)

8.58 D.J. Ehrlich, J.Y. Tsao, C.O. Bozler: J. Vac. Sci. Technol. B 3, 1 (1985)

8.59 J.H. Brannon: in ⌊8.28⌋, p. 112

8.60 K. Daree, W. Kaiser: Glass Technol. 18, 19 (1977)

8.61 J.E. Bjorkholm, A.A. Ballman: Appl. Phys. Lett. 43, 574 (1983)

8.62 D. Moutonnet, S. Mottet, D. Riviere, J.P. Mercier: in ⌊8.1⌋, p. 339; in ⌊8.13⌋

8.63 T.J. Chuang: Surf. Sci. Rep. 3, 1 (1983)

8.64 T.J. Chuang: J. Vac. Sci. Technol. 21, 798 (1982)

8.65 G.L. Loper, M.D. Tabat: Appl. Phys. Lett. 46, 654 (1985)

8.66 G.L. Loper, M.D. Tabat: in ⌊8.56⌋, p. 121

8.67 G.L. Loper, M.D. Tabat: J. Appl. Phys. 58, 3649 (1985)

8.68 R.J. von Gutfeld, R.E. Acosta, L.T. Romankiw: IBM J. Res. Dev. 26, 136 (1982)

8.69 Y. Horiike, M. Sekine, K. Horioka, T. Arikado, M. Nakase, H. Okano: in ⌊8.28⌋, p. 99

8.70 H. Okano, Y. Horiike, M. Sekine: Jpn. J. Appl. Phys. 24, 68 (1985)

8.71 T. Arikado, M. Sekine, H. Okano, Y. Horiike: in ⌊8.8⌋, p. 167

8.72 D.J. Ehrlich, J.Y. Tsao: J. Vac. Sci. Technol. B 1, 969 (1983)

8.73 D.J. Ehrlich, R.M. Osgood, T.F. Deutsch: Appl. Phys. Lett. 38, 1018 (1981)
8.74 L.L. Sveshnikova, V.I. Donin, S.M. Repinskii: Sov. Phys.-Tech. Phys. Lett. 3, 223 (1977)
8.75 W. Holber, G. Rekesten, R.M. Osgood: Appl. Phys. Lett. 46, 201 (1985)
8.76 F.A. Houle: in ⌊8.8⌋, p. 203
8.77 F.A. Houle: J. Chem. Phys. 79, 4237 (1983)
8.78 F.A. Houle: Chem. Phys. Lett. 95, 5 (1983)
8.79 T.J. Chuang: IBM J. Res. Dev. 26, 145 (1982)
8.80 F.A. Houle, T.J. Chuang: J. Vac. Sci. Technol. 20, 790 (1982)
8.81 T.J. Chuang: J. Chem. Phys. 74, 1461 (1981)
8.82 T.J. Chuang: J. Chem. Phys. 74, 1453 (1981)
8.83 A.L. Dalisa, W.K. Zwicker, D.J. DeBitetto, P. Harnack: Appl. Phys. Lett. 17, 208 (1970)
8.84 F.V. Bunkin, B.S. Lukyanchuk, G.A. Shafeev, E.K. Kozlova, A.I. Portniagin, A.A. Yeryomenko, P. Mogyorosi, J.G. Kiss: Appl. Phys. A 37, 117 (1985)
8.85 E.F. Krimmel, A.G.K. Lutsch, R. Swanepoel, J. Brink: Appl. Phys. A 38, 109 (1985)
8.86 J.A. Steinfeld, T.G. Anderson, C. Reiser, D.R. Dehison, L.D. Hartsough, J.R. Hollahan: J. Electrochem. Soc. 127, 514 (1980)
8.87 D. Harradine, F.R. McFeely, B. Roop, J.I. Steinfeld, D. Denison, L. Hartsough, J.R. Hollahan: SPIE 270, 52 (1981)
8.88 R. Solomon, L.F. Mueller: US Patent 3.364.087 (1968)
8.89 W. Sesselmann, E.E. Marinero, T.J. Chuang: Appl. Phys. A 41 (1986)
8.90 J.F. Ready: *Industrial Applications of Lasers* (Academic, New York 1978)
8.91 G. Herziger, E.W. Kreutz: in ⌊8.1⌋, p. 90
8.92 D. Schuöcker (ed.): *Industrial Applications of High Power Lasers,* SPIE, Vol. 455 (1983)
8.93 M. Bass (ed.): *Laser Material Processing* (North-Holland, Amsterdam 1983)
8.94 T.J. Chuang: in *Interfaces under Laser Irradiation,* ed. by D. Bäuerle, L.D. Laude, M. Wautelet, Nato ASI Series (M. Nijhoff, Dordrecht 1987)
8.95 G. Koren: Appl. Phys. A 40, 215 (1986)
8.96 D.V. Podlesnik, H.H. Gilgen, R.M. Osgood: Appl. Phys. Lett. 48, 496 (1986)
8.97 E. Sutcliffe, R. Srinivasan: J. Appl. Phys. (1986)
8.98 T.J. Chuang: private communication
8.99 R. Gauthier, C. Guittard: Phys. Status Solidi A 38, 477 (1976)
8.100 H.F. Winters, J.W. Coburn, T.J. Chuang: J. Vac. Sci. Technol. A 1, 1157 (1983)
8.101 H.F. Winters, J.W. Coburn, T.J. Chuang: J. Vac. Sci. Technol. B 1, 469 (1983)
8.102 F.R. McFeely, J. Morar, G. Landgren, F.J. Himpsel: in *Thin Films and Interfaces II,* ed. by J.E.E. Baghin, D.R. Campbell, W.K. Chu (North-Holland, New York 1984)
8.103 D. Bäuerle: Laser und Optoelektronik 1, 29 (1985)
8.104 J. Otto, R. Stumpe, D. Bäuerle: in ⌊8.1⌋, p. 320
8.105 M. Eyett, D. Bäuerle: unpublished
8.106 M. Eyett, D. Bäuerle: to be published
8.107 JANAF Thermochemical Tables, National Bureau of Standards, USA (1970)
8.108 D.A. Northrop: J. Am. Ceram. Soc. 51, 357 (1968)
8.109 L.M. Kukreja, D.D. Bhawalkar, U.K. Chatterjee, B.L. Gupta: Appl. Phys. A 36, 19 (1985)
8.110 T.F. Deutsch: in ⌊8.1⌋, p. 239
8.111 T.F. Deutsch, M.W. Geis: J. Appl. Phys. 54, 7201 (1983)
8.112 R. Srinivasan, V. Mayne-Banton: Appl. Phys. Lett. 41, 576 (1982)
8.113 R. Srinivasan, W.J. Leigh: J. Am. Chem. Soc. 104, 6784 (1982)
8.114 R. Srinivasan: unpublished

8.115 S. Lazare, R. Srinivasan: J. Chem. Phys. 90 (1986)
8.116 P.E. Dyer, R. Srinivasan: Appl. Phys. Lett. 48, 445 (1986)
8.117 P.E. Dyer, J. Sidhu: J. Appl. Phys. 57, 1420 (1985)
8.118 G. Koren, J.T.C. Yeh: Appl. Phys. Lett. 44, 1112 (1984)
8.119 R. Srinivasan: J. Vac. Sci. Technol. B 1, 923 (1983)
8.120 B. Braren, R. Srinivasan: J. Vac. Sci. Technol. B 3, 913 (1985)
8.121 R. Srinivasan, B. Braren, D.E. Seeger, R.W. Dreyfus: Macromol. 19, 916 (1986)
8.122 R. Srinivasan, B. Braren, R.W. Dreyfus, L. Hadel, D.E. Seeger: J. Opt. Soc. (1986)
8.123 F.N. Goodall, R.A. Moody, W.T. Welford: Opt. Commun. 57, 227 (1986)
8.124 K. Li, M.M. Oprysko: Appl. Phys. Lett. 46, 997 (1985)
8.125 R. Srinivasan: in ⌊8.1⌋, p. 343
8.126 J.H. Brannon, J.R. Lankard, A.I. Baise, F. Burns, J. Kaufman: J. Appl. Phys. 58, 2036 (1985)
8.127 M. Latta, R. Moore, S. Rice, K. Jain: J. Appl. Phys. 56, 586 (1984)
8.128 B.J. Garrison, R. Srinivasan: J. Appl. Phys. 57, 2909 (1985)
8.129 H.H.G. Jellinek, R. Srinivasan: J. Phys. Chem. 88, 3048 (1984)
8.130 R. Srinivasan, P.E. Dyer, B. Braren: to be published
8.134 M. Takai, S. Nagatomo, T. Koizumi, K. Gamo, S. Namba: in ⌊8.13⌋
8.139 V. Srinivasan, M.A. Smrtic, S.V. Babu: J. Appl. Phys. 59, 3861 (1986)
8.140 J.H. Brannon, J.R. Lankard: Appl. Phys. Lett. 48, 1226 (1986)
8.141 G.M. Davis, M.C. Gower, C. Fotakis, T. Efthimiopoulos, P. Argyrakis: Appl. Phys. A 36, 27 (1985)
8.142 R. Srinivasan: in ⌊8.94⌋

Chapter 9

9.1 See, e.g., L.I. Maissel, R. Glang (eds.): *Handbook of Thin Film Technology* (McGraw-Hill, New York 1970)
9.2 For a review see, e.g., J. Bloem, L.J. Giling: in *Current Topics in Materials Science,* Vol. 1, ed. by E. Kaldis (North-Holland, New York 1978) p. 147
9.3 I. Brodie, J. Muray: *The Phyics of Microfabrication* (Plenum, New York 1982)
9.4 N.G. Einspruch (ed.): *VLSI Electronics,* Microstructure Science, Vol. 7 (Academic, New York 1983)
9.5 A.G. Baker, W.C. Morris: Rev. Sci. Instrum. 32, 458 (1961)
9.6 T.J. Chuang, I. Hussla, W. Sesselmann: in *Laser Processing and Diagnostics,* ed. by D. Bäuerle, Springer Ser. Chem. Phys., Vol. 39 (Springer, Berlin, Heidelberg 1984) p. 300
9.7 G.L. Loper, M.D. Tabat: Appl. Phys. Lett. 46, 654 (1985)
9.8 G.L. Loper, M.D. Tabat: SPIE 459, 121 (1984)
9.9 G.L. Loper, M.D. Tabat: J. Appl. Phys. 58, 3649 (1985)
9.10 D.J. Ehrlich, R.M. Osgood, T.F. Deutsch: Appl. Phys. Lett. 38, 1018 (1981)
9.11 T.F. Deutsch, J.C.C. Fan, G.W. Turner, R.L. Chapman, D.J. Ehrlich, R.M. Osgood: Appl. Phys. Lett. 38, 144 (1981)
9.12 J.C. Muller, P. Siffert, C.T. Ho, J.I. Hanoka, F.V. Wald: J. Phys. (Paris) 43, C1-235 (1982)
9.13 E. Fogarassy, R. Stuck, J.J. Grob, P. Siffert: J. Appl. Phys. 52, 1076 (1981)
9.14 R. Stuck, E. Fogarassy, J.C. Muller, M. Hodeau, A. Wattieux, P. Siffert: Appl. Phys. Lett. 38, 715 (1981)
9.15 G.B. Turner, D. Tarrant, G. Pollock, P. Pressley, R. Press: Appl. Phys. Lett. 39, 967 (1981)
9.16 T.F. Deutsch, J.C.C. Fan, D.J. Ehrlich, G.W. Turner, R.L. Chapman, R.P. Gale: Appl. Phys. Lett. 40, 722 (1982)

9.17 T.F. Deutsch, D.J. Ehrlich, D.D. Rathman, D.J. Silversmith, R.M. Osgood: Appl. Phys. Lett. 39, 825 (1981)
9.18 T.F. Deutsch, D.J. Ehrlich, D.D. Rathman, D.J. Silversmith, R.M. Osgood: in *Laser Diagnostics and Photochemical Processing for Semiconductor Devices,* ed. by R.M. Osgood, S.R.J. Brueck, H.R. Schlossberg (North-Holland, New York 1983) p. 225
9.19 C.H.J. v.d. Brekel, PhD Thesis, Eindhoven (1978)
9.20 S. Nishino, H. Honda, J. Matsunami: Jpn. J. Appl. Phys. 25, L87 (1986)
9.21 C. Arnone, M. Rothschild, J.G. Black, D.J. Ehrlich: Appl. Phys. Lett. 48, 1018 (1986)
9.22 C.R. Moylan, T.H. Baum, C.R. Jones: Appl. Phys. A 40, 1 (1986)
9.23 D.A. Doane: Solid State Technol. 23, 101 (1980)
9.24 Karl Suss, Inc.: unpublished
9.25 K. Jain: in Proc. Int. Conf. *Microcircuit Engineering,* Cambridge, U.K. 1983, p. 181
9.26 K. Jain, C.G. Wilson, B.J. Lin: IEEE Electron Device Lett. 3, 53 (1982); IBM J. Res. Dev. 26, 151 (1982)
9.27 K. Jain, R.T. Kerth: Appl. Opt. 23, 648 (1984)
9.28 D.J. Ehrlich, J.Y. Tsao, C.O. Bozler: J. Vac. Sci. Technol. B 3, 1 (1985)
9.29 H. Craighead: private communication
9.30 S.M. Metev, S.K. Savtchenko, K.V. Stamenov, V.P. Veiko, G.A. Kotov, G.D. Shandibina: IEEE J. QE-17, 2004 (1981)
9.31 S.M. Metev, S.K. Savtchenko, K.V. Stamenov, V.P. Veiko, G.A. Kotov, G.D. Shandibina: in *Laser Assisted Modification and Synthesis of Materials,* Proc. Int. Winter School, Gyuletchitza, Bulgaria 1985, ed. by S. Metev, p. 61
9.32 S.M. Metev, S.K. Savtchenko, K.V. Stamenov, V.P. Veiko, G.A. Kotov, G.D. Shandibina: J. Phys. D 13, L 75 (1980)
9.33 R.F. Karlicek, V.M. Donnelly, G.J. Collins: J. Appl. Phys. 53, 1084 (1982)
9.34 A. Kapenieks, M. Eyett, D. Bäuerle: Appl. Phys. A (1986)
9.35 A. Kapenieks, R. Stumpe, M. Eyett, D. Bäuerle: in *Laser Processing and Diagnostics II,* ed. by D. Bäuerle, K.L. Kompa, L.D. Laude (Physique, Les Ulis 1986)
9.36 D. Bäuerle: in ⌊9.6⌋, p. 166
9.37 H.H. Gilgen, C.J. Chen, R. Krchnavek, R.M. Osgood: in ⌊9.6⌋, p. 225
9.38 D.J. Ehrlich, J.Y. Tsao: in ⌊9.6⌋, p. 386
9.39 D.J. Ehrlich, J.Y. Tsao: in ⌊9.4⌋, p. 129
9.40 I.P. Herman: in ⌊9.6⌋, p. 396
9.41 B.M. McWilliams, I.P. Herman, F. Mitlitsky, R.A. Hyde, L.L. Wood: Appl. Phys. Lett. 43, 946 (1983)
9.42 J.Y. Tsao, D.J. Ehrlich, D.J. Silversmith, R.W. Mountain: IEEE Electron Device Lett. 3, 164 (1982)
9.43 J.Y. Tsao, D.J. Ehrlich, D.J. Silversmith, R.W. Mountain: in *Laser Controlled Chemical Processing of Surfaces,* ed. by A.W. Johnson, D.J. Ehrlich, H.R. Schlossberg (North-Holland, New York 1984) p. 55
9.44 T.F. Deutsch: in ⌊9.6⌋, p. 239
9.45 D.J. Ehrlich, R.M. Osgood, D.J. Silversmith, T.F. Deutsch: IEEE Electron Device Lett. 1, 101 (1980)
9.46 W. Kräuter, D. Bäuerle: unpublished
9.47 H.H. Gilgen, D. Podlesnik, C.J. Chen, R.M. Osgood, Jr.: in ⌊9.43⌋, p. 139
9.48 D.V. Podlesnik, H.H. Gilgen, R.M. Osgood, A. Sanchez, V. Daneu: Appl. Phys. Lett. 43, 1083 (1983)
9.49 D.V. Podlesnik, H.H. Gilgen, R.M. Osgood, A. Sanchez, V. Daneu: in ⌊9.18⌋, p. 57
9.50 R.W. Haynes, G.M. Metze, V.G. Kreismanis, L.F. Eastman: Appl. Phys. Lett. 37, 344 (1980)

9.51 A. Yariv, H. Nakamura: IEEE J. QE-13, 233 (1977)
9.52 J.Y. Tsao, R.A. Becker, D.J. Ehrlich, F.J. Leonberger: Appl. Phys. Lett. 42, 559 (1983)
9.53 W. Kräuter, D. Bäuerle, F. Fimberger: Appl. Phys. A 31, 13 (1983)
9.54 F.A. Modine, D.H. Lowndes, J.R. Martinelli, E. Sonder: J. Appl. Phys. 57, 5066 (1985)
9.55 S.P. Murarka: *Silicides for VLSI Applications* (Academic, Orlando 1983)
9.56 S. Metev, S. Savtchenko: in *Interfaces under Laser Irradiation,* ed. by D. Bäuerle, L.D. Laude, M. Wautelet, Nato ASI Series (M. Nijhoff, Dordrecht 1987)

Subject Index

Page numbers printed in italics refer to tables or listings of data or systems

Ablation, laser-induced, *see also* Etching
— inorganic materials 51
— organic polymers 52,162,193,*194*
— — model calculations 52
— — rates *194*
Absorption, optical, *see* Optical
Activation energy
— CH_4 93
— C_2H_2 93
— C_2H_4 93
— C_2H_6 93
— $Ni(CO)_4$ 89
— SiH_4 92
— WF_6 118
Adhesion *70*,132
— Al_2O_3 130
— Au 107
— Si_3N_4 131
— SiO_2 128
Adsorbates 38
— chemisorption 38
— coverage 40
— deposition from *70*,84
— desorption 39,41,83
— doping from *150*,152
— lateral diffusion 40,83
— physisorption 38
— reaction rates 39
Alloys, *see also* Compound formation
— metal 2,157
Annealing
— laser
— — defects 2
— — ion implanted surfaces 2,*150*
— rapid thermal 205
Applications 199
— circuit restructuring 206
— contacts *70*,106,*150*,206
— devices *150*,206
— direct writing 96,206
— electrodes 149,206
— fault correction 206
— integrated circuits 206
— interconnects 206

— junctions *150*,205
— lithography 205
— nonplanar fabrication 91,185,207
— optical waveguides *70*,207
— oxides 126,204
— projection patterning *70*,189,206
— solar cells *70*,120,*150*,205
— 3-D structures 91,186,207
— trimming 206
— VLSI 203
Arrhenius behavior 89,93
Auger process 14

Butler-Volmer equation 48

Cabrera-Mott theory 137
Ceramics
— etching *163*,190
— metallization 146
— reduction 145
Chemical activation energy, *see* Activation energy
Chemisorption 38
Compound formation, *see also* Deposition
— compound semiconductors *159*
— metal silicides *158*
Conductivity
— electrical, *see* Electrical properties
— thermal
— — Ni 18
— — Si 10
— — SiO_2 19
Confinement
— spatial 53
Crystallization
— amorphous films 3
— explosive 116
Cutting, *see also* Ablation, Etching
— laser machining 2,169
CVD 3,42,93,118,200

Deposition, laser-induced 69,*70*, *128,130,131*

241

244

Date Due

Due	Returned	Due	Returned
NOV 14 1988	NOV 31 1988		
APR 14 1989	APR 14 1989		
APR 01 199	APR 17 1996		